Aquaculture Practices in Taiwan

by

T. P. Chen

ISBN 085238 080 1

Printed by Page Bros (Norwich) Ltd

Contents

List of Illustrations

The Author

A lifetime spent in the study and practice of aquaculture in China and Taiwan enables T P Chen to give in this practical and informative work detailed guidance on the successful commercial culture of no fewer than 29 species of fish and one plant. After graduating from college in Peking he studied at the University of Washington, Seattle USA taking his M Sc degree in 1925. On returning to China he speedily won executive positions at important Experiment Stations until interruption occasioned by the war with Japan 1936–1945. He emerged from that period in 1946 as general manager of the Taiwan Fisheries Corporation and from that time onwards he has been in the forefront of leadership in developing aquaculture in Taiwan as is outlined in the introductory chapter of this work. After three year's directorship of the Fisheries Rehabilitation Administration he was for nine years fisheries specialist on the Commission of Rural Reconstruction, finally becoming chief of the Fisheries Division of that Commission for the years 1959–1973. In this period Taiwan's aquaculture made remarkable progress thanks largely to the responsive harmony established between T P Chen and his staff with the interested fishing communities. The present position of the author is chairman of Taiwan Fisheries Consultants Inc.

Throughout his career T P Chen has written a number of important research and scientific papers for many official publications. This work now makes available for the first time the actual practices successfully established in Taiwan so that other areas where it is deemed conditions are suitable and favourable may benefit.

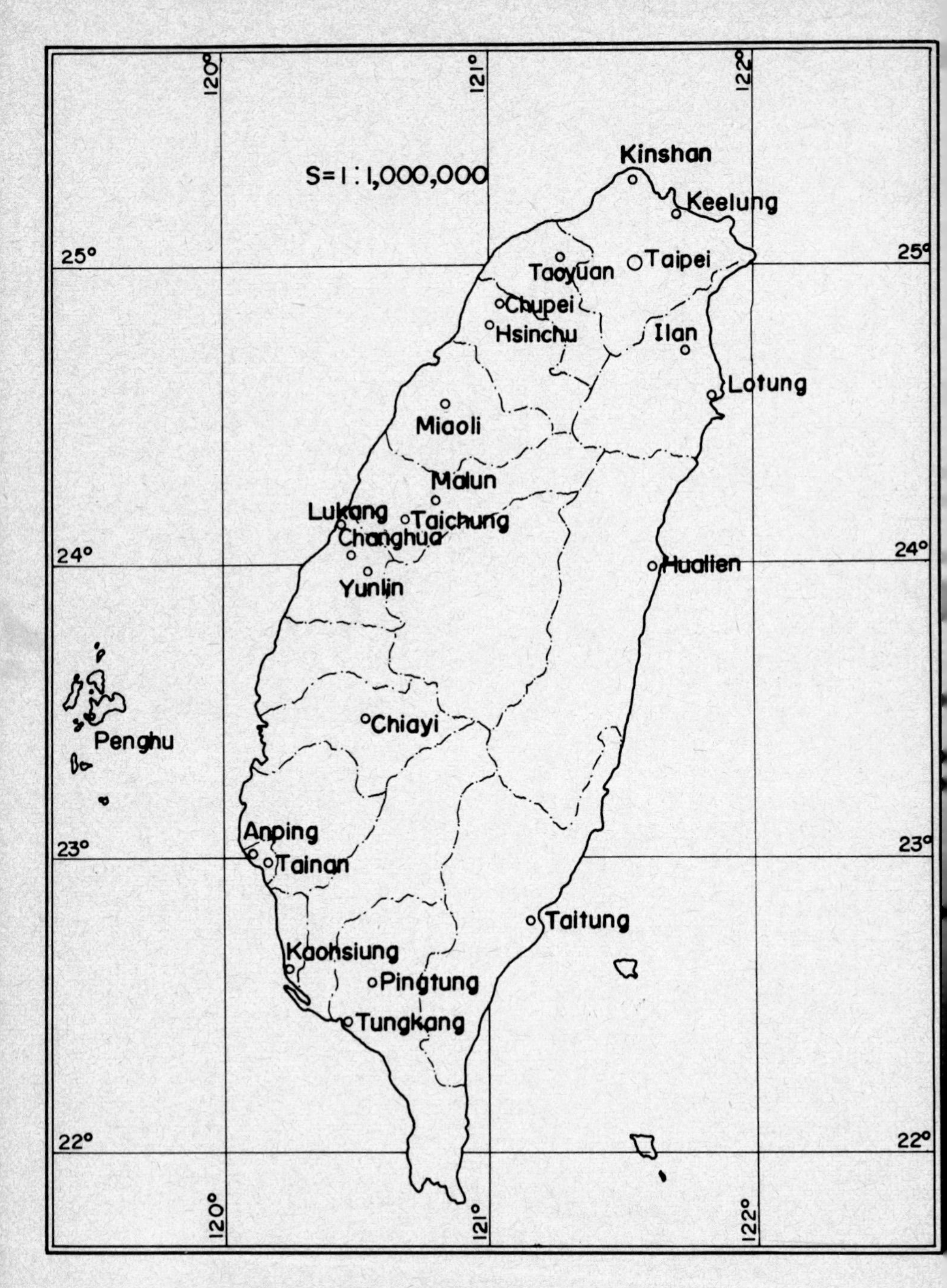

MAP OF TAIWAN AND THE PENGHU ISLANDS

TAIWAN

Taiwan lies some 90 miles east of the Chinese mainland between the north latitudes of 21° 45′ and 25° 38′ and is crossed about mid-length by the Tropic of Capricorn. It is approximately 260 miles long by about 90 miles at the widest part. In area it is 13,000 square miles. The eastern part of the island is mountainous and forest covered. The highest peaks rise to 13,035 feet and 12,072 respectively and the western plains are watered by many rivers from these heights. The soil is very fertile and produces sugar, rice, tea, bananas, pineapples, asparagus and mushrooms. Coal, sulphur, iron and petroleum are also produced and copper and gold are mined. The fisheries of the country are important – both marine and aquaculture – the latter especially having expanded markedly in recent years.

The population exceeds 18,000,000 and is basically Chinese. For many centuries Taiwan was administered as part of China. It was, however, administered as a province of Japan from 1895 to 1945. After the Second World War it was retroceded to the Republic of China which now administers the territory as distinct from mainland China. The two principal seaports are Keelung in the north with a population of about 400,000 and Kaohsiung in the south with a population of about 900,000. The capital Taipei has a population of about 2,000,000. A mutual defence treaty was signed between the United States and Taiwan in 1954.

There is considerable difference between the south and north in the season of rainfall and air temperature. In southern Taiwan the rainy season is in the summer; in northern Taiwan, in the winter. The average monthly temperature in Ilan (the aquaculture centre in northern Taiwan) and Tainan (the aquaculture centre in southern Taiwan) (recorded from 1897 to 1945) are given as follows:

(In centigrade)	Ilan	Tainan
January	15·8	17·0
February	16·0	17·1
March	17·9	19·7
April	20·6	23·4

May	23·5	26·3
June	26·1	27·4
July	27·5	27·8
August	27·2	27·5
September	25·8	27·1
October	22·8	24·8
November	20·2	21·8
December	17·2	18·5

These temperatures are given as a guide to aquaculturists elsewhere in considering the applicability of Taiwan practices in their own localities.

INTRODUCTION

The production from aquaculture in Taiwan (including Penghu) in 1974 totalled 114,472 metric tons, which was 16·4% of the total fisheries production of 697,871 metric tons. The more important species cultured were:

Species	*Quantity of production* (*mt*)	*Value* (*NT$1,000*)
1. Milkfish	28,907	1,067,609
2. Carps	16,816	339,629
3. Tilapia	14,772	220,934
4. Oyster	13,359	559,397
5. Eel	11,827	2,059,513
6. Meretrix clam	11,695	197,911
7. Gracilaria	5,971	21,897
8. Asari clam	2,203	93,358
9. Mullet	1,298	88,191
10. Freshwater clam	1,013	13,382
11. Soft-shell turtle	282	38,715
12. Sand shrimp	223	12,025
13. Serrated crab	150	23,839
14. Grass shrimp	140	21,581
15. Blood cockle	57	11,400

The value of the 108,713 metric tons of produce listed above totalled in Taiwan dollars 4,769,381,000. At exchange rates current at time of preparing this book this sum converted into Sterling at £65,500,000.

It will be seen from this above table that although the eel production ranked fifth in terms of quantity, yet it took first rank in terms of value. It was almost twice the value of milkfish, which had the highest actual production among all cultured species.

Aquaculture in Taiwan is noted for its diversification. In the present book, the cultural practices of 29 species of animals and one species of plant are described. They include one species each of algae, reptiles and amphibians, five species of molluscs, four species of crustaceans and eighteen species of fish. The culture of some of these species is not known or practised in other parts of the world, for example the culture of snakehead, walking catfish, mud skipper, *Corbicula* clam and the viviparous snail. The culture of the seaweed *Gracilaria* in ponds is also unique in Taiwan.

It is the spirit of progressiveness that has been responsible for the success of aquacultural developments in Taiwan. The fish farmers are not only industrious but are eager to acquire new technical know-how. The moment they learn that some new techniques have been developed or some new species have been introduced by the research institutions, they immediately flock to the institutions for information and request demonstrations. There is literally no gap between research and commercial operation. This is markedly different to what is the case in many other areas, where, in spite of demonstrations and government assistance, the fish farmers are often still reluctant to make improvements or adopt new practices. The success in many fields of aquaculture in Taiwan as described in this book will perhaps provide incentives and encouragement to people in these areas.

Taiwan leads the world in some aspects of aquaculture. Examples are the culture of milkfish, the polyculture of Chinese carps and the artificial propagation of the grey mullet. The utilization of sewage for culture of tilapia is also noteworthy. If this can be accomplished in Taiwan, it obviously can be done in other parts of the world where climatic and other conditions are similar to or are even better than in Taiwan. For instance, Philippines and Indonesia, where the climate permits the growth of milkfish all year round, have a combined pond area of over 300,000 hectares used for milkfish farming with a total annual production estimated as 135,000 mt. This production could be increased to about 600,000 mt annually, if only the Taiwan practices were adopted in the existing ponds.

Generally speaking, aquaculture has two purposes: first, the production of food, particularly animal protein food, for the protein hungry population; and secondly, to provide a good income to the fish farmers. The first of these two purposes characterizes the position of aquaculture in Taiwan from 1945 to about 1960. In that period the species produced by aquaculture, predominantly milkfish, carps and oysters, were mostly consumed by the local populace to meet their need for animal protein food. But from 1960 onwards, to meet the demand of the more affluent people, particularly the gourmets, for foods that were new and different in taste, luxury items – such as soft-shell turtles, ayu, rainbow trout – were produced in even greater quantity by the fish farmers. An export trade too began to expand and it provided an incentive to fish farmers to grow certain species such as the eel. The export of eel to Japan in 1974, about 7,000 mt was valued at about

US$ 50 million. This was accompanied by the steady increase over the years of eel farm acreage to about 1,200 ha (or about 3,000 acres).

During the 14 years that the author has served as Chief of the Fisheries Division, Joint Commission on Rural Reconstruction in Taipei, many aquaculture workers from other lands have come to Taiwan for training in aquaculture, including some from Thailand, Malaysia, South Vietnam, Philippines, Indonesia, Palau, Korea and Nigeria. It is hoped that this book will benefit not only people in these countries but all others who are interested in acquiring information on aquaculture as practised in Taiwan and so be able to apply the techniques developed here to their own lands.

It should be stressed, however, that many of the practices as described in this book cannot be adopted directly as such in other areas. This is due mainly to environmental, institutional and socio-economic differences, such as temperature, water supply, land availability, cost of feeds and labour and market demand. The aquaculturists in other areas will have to use their judgement in adopting and modifying these practices.

It has taken the author about two years to prepare this book. Thanks are due to Mr Ting-lang Huang and Dr I C Liao for supplying information on some of the subjects and to Mrs Ruth Lee and Miss Gloria Luh for assistance in the preparation of the manuscript.

T P Chen

I. Milkfish Farming

1. INTRODUCTION

The milkfish, *Chanos chanos* (Forskal) (Fig 1), has a geographical distribution from longitude 40°E to about 100°W and from latitude 30–40°N to 30–40°S. This area covers the tropic and subtropic sections of the Indian and the Pacific Oceans.[1] It has not been reported from the Pacific coast of the United States north of California.

Farming of milkfish is an important industry in the Philippines, Indonesia and Taiwan. Although farming the smallest acreage among the three regions, Taiwan has the highest yield per unit area. According to the Fisheries Yearbook published by the Taiwan Fisheries Bureau,[2] the total milkfish culture area in 1972 was 15,624 ha with a total production of 24,950 mt. The average yield per hectare of pond was therefore about 1,600 kg. This includes yield figures from newly developed ponds which have low production. Yields as high as 3,000 kg/ha are reported from some of the good ponds as a result of adequate fertilization and stock manipulation.

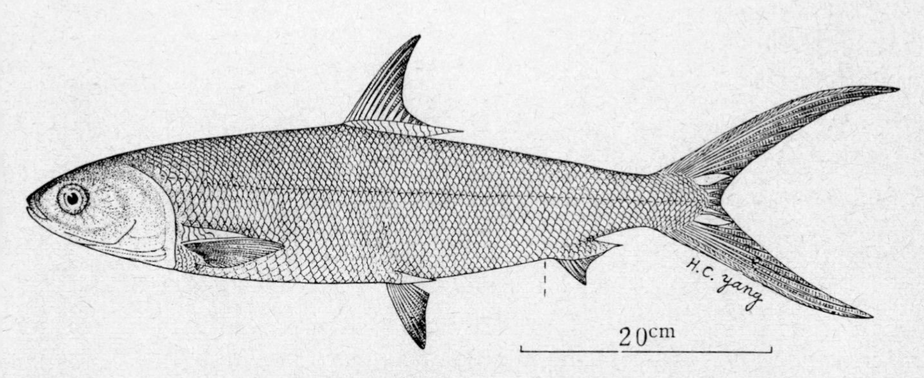

Fig 1 Milkfish *Chanos chanos*

2. HISTORY

It is generally believed that the culture of milkfish in Taiwan dates back to the reign of General Cheng Cheng-kung (Koxinga) more than three hundred years ago.[3] The Chinese fish farmers who migrated to Taiwan built dykes on the low land along the coast of Taiwan and stocked the ponds so built with the fry of milkfish obtained from the littoral waters. Milkfish ponds were first built in the Anping area of Tainan.

About 1910, during the Japanese occupation, a fish culture station was established in Tainan by the Taiwan Governor-General with Takeo Aoki as its head to carry out experiments in milkfish culture.[4] This station, which became a field station of the Taiwan Fisheries Research Institute in 1945, is now being moved to a new site of 70 ha in the Tsengwen Tidal Land area.

3. SUPPLY OF MILKFISH FRY

Estimated at 10,000 per hectare, the number of milkfish fry required annually for stocking the 15,624 ha of milkfish ponds in Taiwan is about 160,000,000. In recent years, due to the increase of ponds for production of smaller milkfish as longlining bait, which practice requires heavier stocking, the demand for fry has increased considerably. On the other hand, the production of milkfish fry fluctuates widely from year to year (Table 1), resulting often in shortages.

Table 1. Yearly Fluctuations of Production of Milkfish Fry, 1963–1972

Year	Production	Year	Production
1963	85,000,000	1968	96,562,000
1964	153,200,000	1969	137,456,000
1965	98,200,000	1970	207,000,000
1966	187,000,000	1971	133,250,000
1967	28,000,000	1972	128,310,000

The milkfish fry are captured in Taiwan in the months from April to August,[3] with the peak during April to June. The maximum catch occurs at full or new moon, ie at the time of the spring tide. During these months, thousands of coastal people, women and children included, gather on the beaches and in estuaries and coves to reap the harvest of the sea.

The gear most commonly used is a triangular scoop net (Fig 2), which is pushed forward by the fishermen wading in chest high water. A drag net is also used at some places. It is towed by two ropes, one of which is attached to a bamboo raft and the other is towed by a man who walks along the shore.

At shallow places, the drag net can also be towed by two persons, each holding onto the tip of each wing.

In 1967, the fry collectors in the southern part of the island began using powered sampans to catch the milkfish fry. Two nets were towed, one on each side of the boat. These boats can operate as far as one kilometre out from the shore and they catch many more fry than is done by the older methods.

The milkfish fry captured along the coast of Taiwan are free from yolk sacs. They are transparent and average 1·5 cm in length.

Fig 2 Scoop net for catching milkfish fry

In some years, when the local supply of milkfish fry was not sufficient, quantities were imported from the Philippines and Indonesia. Because these countries are often plagued by cholera, the government of Taiwan has recently required that the milkfish fry be sterilized with Panfuran-S in concentration of 0·5 ppm upon their arrival.

4. TRANSPORTATION OF MILKFISH FRY

After their capture, the milkfish fry are kept in sea water in wooden buckets or are temporarily put in small cement troughs on the beach. They are sold to fry collectors who take them to Tainan and sell them to milkfish fry dealers. Formerly, the fry were put into water-tight baskets or galvanized iron cans with sea water and transported by train or automobile. Since the trip required four to ten hours, two changes of water and its agitation were necessary for the longer trips. Since 1962, the use of plastic bags has become prevalent. The bag contains water of 10–15‰ salinity and is filled with oxygen. Ice is added during the hot weather. Thus mortality is lowered and much labour saved.

5. MILKFISH FRY DEALERS

The milkfish fry dealers are concentrated in the city of Tainan. There are seven dealers at the time of writing. Each dealer has a number of small cement tanks of about ten square feet in area and about seven inches in depth of water. Each tank holds about 30,000 fry. As the fish fry dealers are located in the city, tap water is often used. Salt is added to obtain a salinity of 5 to 12‰.

The fry are fed with wheat flour (occasionally egg yolk) twice a day. About an hour after feeding, one bucket of fresh water is added to each tank. This has the effect of starting the circular swimming motion of the fry and the excretion of faeces. This is removed with a dipper from the centre of the bottom after the water is stirred and begins a clockwise circular motion. Water is changed once or more daily. Fish fry remaining too long in the tank may become stunted by the increase of salinity to about 25‰.

The fish fry dealer sells the fry to fish farmers who come to the city to buy. The price of the fry fluctuates widely from day to day according to supply and demand.

6. MANAGEMENT OF MILKFISH PONDS

6.1 Area and Climate

The milkfish ponds in Taiwan are confined to the coastal areas of Yunlin, Chiayi, Tainan, Kaohsiung and Pingtung, where climatic conditions permit a growing season of about eight months per year. No milkfish ponds exist in the area north of Yunlin, which has a longer and colder winter, and on the east coast, which has no tidal flats.

6.2 Location

The ideal location of a milkfish pond is one in which (1) the temperature is above 15°C in about eight months of the year, (2) there is no great exposure to danger of flooding, (3) the pond water is not liable to serious dilution during the rainy season, and (4) fresh water is available to adjust the pond water to a salinity of not over 50‰ during the dry season.

6.3 Elevation

The optimum elevation of a fish pond is from the mean sea-level to an altitude of 0·45 m.[5] The average tidal difference on the west coast of Taiwan is less than 1 m. Full drainage of the pond is difficult if the pond bottom is below the mean sea-level. Such ponds are found in the Pingtung area, where milkfish are reared in association with shrimp and crab. On the other hand, if the elevation is more than 0·45 m, then pumping has to be used to fill the pond. Most of the ponds in the Taiwan area have an elevation of 10–30 cm above mean sea-level.

6.4 Soils

According to Tang and Chen,[5] the algal pasture soils of milkfish ponds in Taiwan consist of alluvial tidal flats of recent origin. Silty loam and loamy soils are most favourable for growing pasture algae. Seepage is small with this type of soil, and it forms strong dykes. It should contain suitable amounts of organic matter to serve as nutrient for the benthic algae. This accounts for the fact that newly constructed milkfish ponds are of low productivity.

6.5 General layout of ponds

A unit for milkfish production consists of the following: dykes (outer and inner), rearing ponds, wintering ponds, nursery ponds, water passage (for filling and drainage) and sluice gates (outer and inner).

The purpose of the outer dyke, or sea dyke, is to resist the encroachment of seawater. It has to be high enough to prevent appreciable overflow of seawater over the top and structurally strong enough to remain stable under the static pressure and dynamic attack of tide and wave. The designer must know the height of the highest sea water level, the characteristics of the highest wave which may occur at the dyke site and the run-up height of wave on the dyke.

On the western coast of Taiwan, the slope of the outside portion of the sea dyke is generally 1:2·5 to 1:3, and the inner portion is 1:2. The top of the dyke is about 2 m in width.

The inner dykes are built for the purpose of dividing the milkfish ponds according to ownership and are generally 1·7 m in height with a slope varying from 1:1·5 to 1:2 according to the nature of the soil. The top is about 4 m in width to facilitate transportation by truck.

In addition to the above, the ponds are further divided by small dykes of 0·8 to 1·0 m in height. The inner surface of these small dykes is generally lined with bricks laid on loosely over the soil. This has been found to be quite economical, because the damaged bricks can be removed and replaced very easily.

The water passage or canal serves as a passage for incoming and outgoing water, for bamboo rafts used to transport feeds, fertilizers, fish, etc, and for temporarily holding the fish when needed. There are inner water passages and outer water passages. The former are generally 6 m in width and 1·5 m in depth. The width of the outer water passages varies according to the size of the ponds they serve and the tidal condition. For ponds of 20 ha in total area, their width is generally 12–15 m. The water passages should have a lower elevation than the ponds they serve to facilitate drainage.

Sluice gates are usually placed at points where the elevation is lowest. The sluice gate at the outer dyke is made of bricks. The series of wooden boards, which are inserted into grooves on both sides of the gate, can be removed or installed one by one to admit or discharge water in the desired quantity or level. Inner sluice gates are much smaller and formed by wooden boards on both sides of the inner dyke.

The shallow pond in front of the wintering pond is often used as a nursery pond. Water in the nursery pond is maintained at a depth of 20–50 cm.

Wintering ponds (Fig 3) are often located close to the outer water passage where the admission and the drainage of water are easy. They are long, deep ditches protected on the windward side by windbreaks, which consist of bamboo frames thatched with straw or other plant material. The width of the wintering pond is about 5 m at the top and about 1·5 m on the bottom. The depth is about 2 m. The windbreak is placed obliquely at an angle varying from 25 to 40 degrees to ward off the wind.

The rearing ponds are generally 3–6 hectares in area. They are rectangular in shape, with the longer axis running from east to west, in order to minimize disturbance of the water from the monsoon wind from the north. The narrow width also facilitates harvesting of the fish by seining. The pond bottom should be levelled, with shallow trenches leading from the water inlet to the outlet.

6.6 Rearing of the fingerlings and nursery ponds

Two kinds of milkfish fingerlings are used for stocking the ponds, the overwintered fingerlings and the new fry captured from the sea in spring.

The overwintered fingerlings are either those undersized (less than 150 g) fish harvested by the fish farmers at the close of the rearing season and kept in wintering ponds, or those late fry captured from the sea in July and August and stocked in wintering ponds by fish farmers who specialize in this trade. In the former case, the undersized fish are placed in the wintering ponds as described above. In the latter case, the facilities consist of shallow ponds (20–40 cm) and deep wintering ditches (1·5 m). Feeds are generally rice bran, peanut meal and soybean meal. Benthic algae also serve as feed during the initial period. To stunt the growth of the fingerlings, as many as 300,000 to 500,000 fingerlings are planted to each hectare.

Fig 3 Milkfish wintering pond

The new fry purchased from the fry dealers are planted in the nursery pond, which is a small pond of less than 20 cm in water depth and 100 to 200 m^2 in area. This is often a temporary structure. When its embankment is removed, it becomes a part of the rearing pond after the fingerlings are transferred to the rearing pond. The usual practice is to utilize the shallow area of the wintering ponds for this purpose. The water in the nursery pond is diluted to 10–15‰ before the fry are introduced.

After the milkfish are held in the nursery pond for a day or less, an opening is made between the nursery pond and the rearing pond. This opening is screened to prevent the fry from entering the rearing pond. The interchange of water results in the water in the two ponds approaching the same salinity. The screen is then removed and the fry are allowed to swim into the rearing pond of their own accord.

6.7 Preparation of the rearing pond

Since the temperature from November to March is too low for the growth of milkfish, the rearing ponds in this period are treated to promote the growth of benthic algae, which are the main food of the milkfish. They consist of blue-green algae (*Oscillatoria, Lyngbya, Phormidium, Spirulina, Micrococcus,* etc) and diatoms (*Navicula, Pleurosigma, Mastogloia, Stauroneis, Amphora, Nitzschia,* etc).[6] A good growth of benthic algae is responsible for the success of milkfish farming. The following is the practice in the Tainan area.

In the middle of November, the water in the pond is drained off, the bottom levelled and the dykes are repaired. The bottom is then exposed to the sun until the soil cracks and crumbles easily under the pressure of the finger (Fig 4). Seawater is then let in to a depth of 10–15 cm. If the pond is deficient in organic matter, the pond should be fertilized with dried chicken droppings or hog manure (500–1,000 kg/ha) or rice bran (300–500 kg/ha) before the introduction of sea water. More fertilizers should be used on newly constructed ponds. The sea water is allowed to evaporate until the bottom is dry, which will be the end of December or January. Sea water is again let in in February. Before the introduction of sea water a second time, the bottom should be fertilized again if the growth of algal bed is found to be poor. With adequate application of fertilizers and the accumulation of nutrients from the introduction of sea water, the algal beds should have developed to a thickness of 1–3 mm by March, and sea water is let in to a depth of 10–15 cm. At this time, tea-seed meal is often added to eradicate the undesirable species of fish in the pond. The cakes of tea-seed meal are simply broken up into small pieces and cast into the water at various points.

6.8 Stocking

Stock manipulation is the key to the high yield of milkfish ponds in Taiwan. Its main objective is the maximum utilization of the food resources (in this case benthic algae) in the pond.[7] In Taiwan, this is accomplished by the following management technique:

In the early part of April, the pond is stocked with overwintered fingerlings of size varying from 5 to 150 g in weight. The number of fingerlings planted varies from 3,000 to 5,000 according to size.

Fig 4 Pond bottom is sun-dried till it cracks

Beginning from April, new fry collected from the sea become available. They are purchased and planted at intervals of two to four weeks. Each time about 1,500 are planted. Table 2 shows the number planted at different times of the year.

Table 2. Number of Fingerlings Planted in Different Months

Month of stocking	*Average weight of fingerlings planted*	*Number of fingerlings planted*
April	5–150 g	5,000
May	0·05	2,500
June	0·06	2,500
July	0·06	2,000
August–September	0·06	3,000
Total number planted		15,000

6.9 Harvesting

Harvesting of the fish begins at the end of May, and about eight harvests are made until the middle of November. The overwintered fingerlings grow to a weight of 300–400 g, at which size they are harvested at about four different times from the end of May to the middle of August. The new fry grow to a weight of 200–300 g in August or September, when they are harvested. The new fry planted in July do not reach marketable size by November and are put into the wintering ponds.

Fig 5 Gill nets for harvesting the crop

The highest total weight of fish in the ponds occurs in June and July, when it reaches 700–800 kg/ha. This is perhaps the highest carrying capacity, at which point the fish population should be thinned out by frequent partial cropping as mentioned above.

Gill nets are used for harvesting the milkfish (Fig 5). The mesh size of the net varies with the size of the fish to be harvested. The height of the net is about 1·6 m. In operation, a number of nets, each about 30 m in length, are joined together and towed across the length of the pond.

Before actually catching the fish, a net of large mesh size or a device consisting of a rope to which a number of bamboo pieces are strung is towed across the pond to scare the fish into emptying their stomachs. Sometimes a number of persons on moving bamboo rafts strike the water with bamboo poles to accomplish the same purpose. The milkfish with empty stomachs keep better when they are trucked to the market.

When the fish are harvested at the end of the rearing season in October or November, the ponds are drained of water after netting, and any fish left over are picked up.

The fish are packed in bamboo baskets (Fig 6) loaded onto trucks and taken to the market. Crushed ice is spread on top of each basket to preserve freshness.

The ordinary size of fish harvested for food is from three to four fish per kilogramme. Those used as bait for longlining are harvested at the size of 10 to 12 fish per kilogramme. As the milkfish grow to over 10 kilogrammes in nature, these fish are indeed babies.

6.10 Feeding and fertilization

Beginning in May when the rainy season starts in south Taiwan and the salinity of the pond water becomes too low to support a good growth of benthic algae, the milkfish in the ponds are given supplementary feed. Prior to this, little or no artificial feeds are given, because the benthic algae are quite abundant in the ponds and some of the organic fertilizers applied in March may still remain to serve as food.

The feeds commonly used are rice bran, soybean meal and peanut meal. The quantities given are generally 30 kg of rice bran or 25 kg of soybean or peanut meal per hectare daily.

Fertilizers in the form of chicken and hog manure are also applied daily during a clear day to maintain the growth of the algal beds. No feeds or fertilizers should be given on cloudy or rainy days. Otherwise excessive production of phytoplankton will result.

6.11 Pests and predators

The common pests in a milkfish pond are the Chironomid larvae, the polychaete worm *Nareis* and the salt-water snail *Corithida,* of which the Chironomid larvae are the most important.[8] It is estimated that during the early part of summer when these larvae attain the quantity of 200 to 500 kg/ha, the benthic algae they consume daily may amount to 30 to 60

Fig 6 Fish are carefully packed in bamboo baskets

Fig 7 Pesticides are applied

kg/ha. They also consume other organic matters on the pond bottom and thus lower the fertility of the pond.[9]

The Chironomid larvae on the pond bottom may be eradicated by applying Accothion, Abate, Lebacid, Sumithion or some other pesticide,[10] without harming the milkfish (Fig 7).

6.12 Wintering

The milkfish being a semi-tropical fish is sensitive to low temperature. According to Yamamura,[11] it becomes sluggish when the water temperature drops below 15°C and dies at a temperature of about 12°C. A great deal, however, depends on the environment and the condition of the fish. It may die at a higher temperature if conditions are unfavourable or if the exposure to cold is prolonged.

As the temperature of pond water may drop to 5 or 6°C in the winter in southern Taiwan, the milkfish could not survive the winter unless they are protected. It is customary to hold some of the milkfish over the winter, because (1) the late fry planted after June do not reach marketable size in October and (2) some fish should be made available for stocking the ponds early in the next rearing season as the 'new fry' will not be available until April.

The milkfish are placed in the wintering ponds in October or November and held until March, when they could be planted in the rearing ponds. There is usually a shallow area adjacent to the wintering pond in which the milkfish can graze and be fed (with rice bran) on warm days. On really cold days, the milkfish do not eat.

Intensive care is necessary to reduce mortality in the wintering ponds. The causes of high mortality are (1) low temperature (below 11°C), (2) oxygen deficiency and (3) accumulation of waste materials on the bottom and consequent production of noxious substances.[6]

As winter is the drought season in southern Taiwan, the water in the wintering ponds tends to gain salinity due to evaporation. It is often necessary, therefore, to add fresh water to the ponds to maintain the proper salinity.

7. PROBLEMS

The main problems in milkfish farming in Taiwan are: (1) the often inadequate supply of fry and (2) high mortality in the wintering ponds.

As mentioned before, milkfish fry are obtained from the sea. They are abundant in the Philippines and Indonesia, but often deficient in Taiwan.

The problem is aggravated by the recent practice of rearing milkfish as bait for longlining which requires heavy stocking at short intervals. To solve the problem, the Taiwan Fisheries Research Institute, with the assistance of the Joint Commission on Rural Reconstruction (JCRR), has begun an experiment to produce milkfish fry by induced spawning in its Tungkang Marine Laboratory. Thus far difficulty has been encountered in obtaining ripe males and gravid females of good condition at the same time. Steps are being taken to improve the facilities.

Shortage of stocking material is at times the result of high mortality in the wintering ponds. When mortality ran as high as 60% in January and February of 1974, there was a serious shortage of overwintered fish for stocking the ponds in March.

To prevent the temperature of the water in the wintering ponds from dropping to a dangerous level, the Tainan Fish Culture Station experimented on covering the pond with a plastic-sheet tent with ventilation holes and found that the water temperature in the covered pond could be maintained at 2°C higher than in the uncovered pond. However, this is not sufficient protection during a severe cold spell and the cost is high. The solution to this problem seems to be the use of underground water, which has a temperature of not less than 20°C in the winter to raise the temperature in the wintering ponds. To encourage this practice, JCRR has subsidized the digging of a number of deep wells in the Tainan area for the purpose of demonstration. The only drawback is that the water from some wells often has high iron content.

Recently (since 1974), several devices have been tried to raise the temperature of the water in wintering ponds. The most effective one consists of a propane gas water heater to supply recirculated hot water to the pond by pipes. This has been found to be very convenient, and since the cold spell normally last for only a few days, it is not too expensive. Plastic beads, the size of marbles, are used to cover the water surface to reduce heat loss.

LITERATURE CITED

1. Schuster, W. H.: Synopsis of Biological Data on Milkfish. *Indo-Pacific Fisheries Council Occasional Paper 59/3.* 1960.
2. Taiwan Fisheries Bureau: *Taiwan Fisheries Yearbook, 1972.*
3. Chen, Tung-pai: Milkfish Culture in Taiwan. *JCRR Fisheries Series No. 1.* 1952.
4. Yamamura, Makio: Lecture on Milkfish (unpublished).
5. Tang, Yun-an and Shing-Hsiang Chen: A Survey of the Algal Pasture Soils of Milkfish Ponds in Taiwan. *Proceedings of the FAO World Symposium on Warm-water Pond Fish Culture,* 1966.
6. Ling, S. W.: Feeds and Feeding of Warm-water Fishes in Asia and the Far East, *Proceedings of the FAO World Symposium on Warm-water Pond Fish Culture,* 1966.
7. Tang, Y. A.: Stock Manipulation of Coastal Fish Farms. *Coastal Aquaculture in the Indo-Pacific Region,* 1972.
8. Tang, Y. A.: Improvement of Milkfish Culture in the Philippines. *IPFC Current Affairs Bulletin No. 49,* August 1967.
9. Tang, Y. A. and T. P. Chen: Control of Chironomid Larvae in Milkfish Ponds. *JCRR Fisheries Series No. 4,* June 1959.
10. Tsai, S. C. and T. L. Huang: Control of Chironomid Larvae in Milkfish Ponds with Abate 50% E.C. *JCRR Fisheries Series No. 8.* 1969.
11. Yamamura, Makio: Milkfish of Taiwan (unpublished).

II. Eel Farming

1. INTRODUCTION

The culture of eel in Taiwan has a comparatively short history. Before the retrocession of Taiwan to the Republic of China, experimental rearing of the Japanese eel, *Anguilla japonica*, was carried out by some of the government fish culture stations. Until 1952 there was no eel culture on a commercial scale. This was due to the small demand for eel and, consequently, the lack of interest on the part of fish farmers. Around 1950 the chief demand for eel was as live baits for shark longlining.

In 1952, an eel farm was established at Taoyuan by the government-operated China Fisheries Corporation. The Taiwan Fisheries Research Institute also started experimental eel farming at Lukang in 1956. These attempts were followed by eel farming on a small commercial scale. In 1966, there were only about 60 ha of eel ponds in Taiwan; in 1967, the area of eel ponds increased to 80 ha. Since 1968, as a result of heavy demand from Japan for seed eels and later for eels of table size, the eel farms have increased rapidly in number as well as in size. The total acreage reached 660 ha in December 1971 and 1,058 ha in June 1972. The value of export of eels to Japan in 1971 exceeded US$10 million and was reported to be US$29·5 million in 1972.

In spite of the continued heavy demand and the high price paid for eel by Japan, the eel farming industry in Taiwan is plagued by one serious problem – the inadequate supply of seed eel, which will be discussed in detail in the accounts that follow.

2. SPECIES CULTURED

The eel cultured in Taiwan consist mainly of two species, the Japanese eel, *Anguilla japonica* (Fig 8), which is a native of the Taiwan area, and the European eel, *A. anguilla*, which is imported as seed eel in recent years.

Another exotic eel, *A. marmorata*, known locally as 'donkey eel', has a high market value, but an attempt to culture it did not succeed.

As the family Anguillidae has only one genus, all eels of this family belong to the genus *Anguilla*. Some other better known species are:

A. diefenbacker (New Zealand)
A. australis (New Zealand)
A. bicolor (Indonesia, Philippines)
A. pacificus (Indonesia, Philippines)
A. rostrata (U.S.A.)

The life history of the *A. anguilla* and *A. rostrata* is largely known. They both grow from elvers and mature in rivers and migrate downstream into the ocean to spawn in the Sargasso Sea, but very little is known of the life history of the *A. japonica*, particularly its spawning ground.

3. POND AREA AND DISTRIBUTION

According to a survey made by Kuo, Huang and Su,[1] the areas of eel ponds in Taiwan in January 1972 were distributed as follows:

	Rearing Ponds	Nursery Ponds*	Total
Changhua County	212·30 ha	—	212·30 ha
Pingtung County	198·19 ha	10·00	208·19 ha
Ilan County	49·03 ha	68·15	117·18 ha
Yunlin County	69·81 ha	23·17	92·98 ha
Chiayi County	43·84 ha	20·00	63·84 ha
Taoyuan County	44·22 ha	10·98	55·20 ha
Hsinchu County	16·09 ha	6·91	23·00 ha
Other Counties	50·38 ha	9·05	59·43 ha
Total	683·86 ha	148·26	832·12 ha

* Ponds used for rearing the glass eel to stocking size. Some eel farmers specialize in rearing the glass eel to various sizes and selling them to others to grow to market size.

It is estimated that the total eel farming acreage had reached 1,200 ha by the middle of 1973. Changhua, Pingtung, Yunlin and Ilan Counties continue to be the main areas of eel farming. In Ilan, irrigation water is used in most cases for providing running water to the ponds; whereas in the three other areas underground water is used.

4. POND CONSTRUCTION AND WATER SUPPLY

Eel ponds require plenty of fresh water and should be located in areas of good water supply both quantitatively and qualitatively. The water should be free from pollution and have pH value between 6·5 and 8·0. The Japanese eel can live both in tropical and temperate environment, but they generally cease to feed and grow at a temperature below 12°C.

In Taiwan, most of the eel ponds use underground water from deep wells, but irrigation water is also used in some parts of northeastern Taiwan. Irrigation water is not desirable, because it may be polluted, particularly with pesticides. Underground water should be aerated by spraying or agitation.

Most eel ponds have perpendicular concrete or brick walls and sandy soil bottoms (Fig 9), although some are mud ponds with steep mud embankments. It is claimed that if the water quality is good and the eels well fed, they will not try to escape even in ordinary fish ponds built of mud.

Some eel ponds have concrete protruding lips at the top of the concrete walls. These are now considered to be unnecessary, except for seed eel rearing ponds. The seed eel, being light in weight, can climb considerable distances up a perpendicular wall. Escape is made usually during heavy rain.

Most ponds are equipped with paddle wheels driven by electric motors (Fig 10). The whirling wheels with paddles turn round and round on the water surface to act as aerators. They are turned on usually at night and in the early morning when the dissolved oxygen is low.

After the practice of Japanese eel farmers, some eel ponds have the portion equipped with agitators partitioned off into a 'pool', where the eels can congregate at time of low oxygen supply. This is, however, not common.

Some eel farmers install a compressed air blower at the bottom of the pond (Fig 11) to increase the oxygen supply. Again, this is uncommon.

Most eel ponds are now constructed above ground (with the bottoms on ground level) to facilitate drainage and save the cost of excavation. But ponds that use irrigation water have to be built below the level of the irrigation canals.

An eel farm requires a large number of ponds, because eels of different sizes have to be segregated to maintain more uniform growth. The average size of a pond is 0·1 to 0·2 ha.[3] Large-size ponds of over a hectare in area may also be found. Ponds for running water eel farming, however, are generally small.

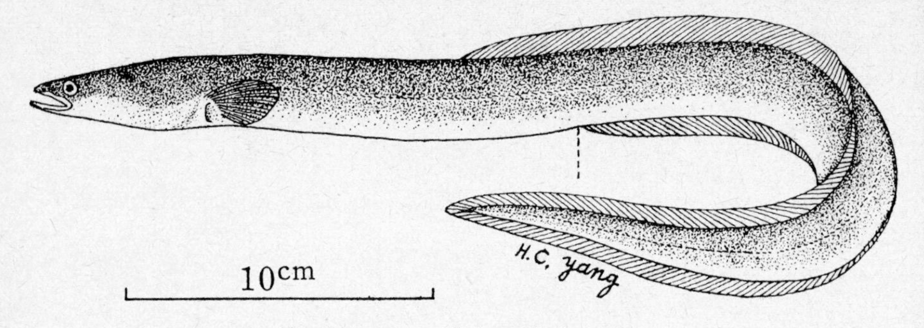

Fig 8 Japanese eel *Anguilla japonica*

Fig 9 Eel ponds in Lotung

Fig 10 Paddle wheels driven by electric motors

5. SUPPLY OF SEED EEL

The natural history of the Japanese eel is not fully known in spite of investigations made by Japanese scientists. It is generally believed that the adult eel, after spending some years to grow and mature in fresh water, descend the rivers to the sea to spawn at some locality south of Taiwan.[2] After hatching, the larval eel, known as leptocephalus, follow the Kuroshio current and drift into estuarine waters, where they are caught. By that time, the leptocephalus have transformed to become a transparent young eel of several centimetres in length known as glass eel. This long and complicated life history explains why it is yet impossible to spawn the eel artificially and rear the larvae to glass eel size.

The seed eel catching season in Taiwan is from October to March, with the peak in December and January.[1] The glass eel of another local species, *Anguilla marmorata*, are found in abundance along the east coast, but no success has been obtained in rearing them in ponds. It is said that, in spite of feeding, they increase very little in size.

The glass eel are caught with scoop nets pushed by one person wading in shallow water, by drag nets towed by a powered raft, or by trap nets set in the estuaries.[1] The glass eel come in with the tide, usually at night.

The glass eel caught are temporarily placed with water in plastic containers to wait for buyers. The buyers, after collecting sufficient quantities of glass eel, sell them either to the eel farmers or middlemen.

No other commodity has experienced a rise in price within the span of a few years as fabulous as the seed eel in Taiwan. Before the commercialization of eel farming, the seed eel had no market and were caught as duck feed. Later, in about 1966, the price for one kilogram of glass eel (*A. japonica*) was only NT$60 (NT$40 = US$1·00). The current market price (March 1973) is now about NT$60,000, an increase of 1,000 times in seven years.[2]

This phenomenal price increase of the seed eel is due to two reasons, the shortage of supply and the heavy buying and high price paid by the Japanese eel importers.

6. TRAINING OF THE ELVERS

The glass eel of *Anguilla japonica* are caught in estuarine waters all along the coast of Taiwan. They are transparent and about 5–6 cm in length, and about 6,000 individuals make up each kilogramme. As they ascend the rivers, the elvers assume a dark colour and are difficult to catch.

In the pond, the elvers have to be trained to take artificial feeds. They are fed with tubifex worms or minced meat of oyster and clam. Minced fish of good quality may also be given. The feeds in a basket are lowered to the bottom of the pond at night. At first the elvers are reluctant to eat, and left-over feeds must be replaced with fresh feeds each night. This is kept up for about 20 days, when most of the elvers will come to feed. Gradually, the feeding time is changed to early morning, and the feed basket is raised to just below the surface. At the same time, minced trash fish or mixed eel feed is gradually added to the feed, until finally the feed consists entirely of minced trash fish or commercial eel food. From then on, the elvers will grow rapidly.

Some eel farmers treat the tubifex with 0·2% sulfamonomethoxine solution for one hour and wash them before giving to the elvers.

7. FEEDS AND FEEDING

The eels feed mainly on animal protein food. During the initial stage of eel farming in Taiwan, silkworm pupae formed the main feed of the eel. Because of the decline of the silk industry and the expansion of the eel farming area, the supply of pupae has become inadequate, and trash fish (small fish of poor quality from trawlers) and scraps from fish processing plants have been used extensively as eel feeds.

The trash fish or scraps are minced with a meat grinder to form a thick paste (Fig 12), which is then placed in a wire basket and lowered to a point just below the surface of the eel pond. At a few eel farms, the trash fish or scraps are cooked before mincing. The merit of this practice is yet to be determined.

Most eel farms are equipped with a small feed processing plant with powered grinders, and large farms have cold storage facilities as well.

Every eel pond is provided with a feeding platform from which the feed basket is lowered. Feeding time is early in the morning. The eels in the pond, having learned when and where they can get food, will come quickly to the wire basket, enter through the mesh, and literally bury themselves in the loaf of, say, fish paste, eating very voraciously.

The Taiwan Fisheries Research Institute began experiments on the formulation of an artificial feed for eel in 1966. In 1968, it produced a mixed feed in powder form with the following formula:

	%
White fish meal	61·0
Gamma starch	14·0
Defatted soybean powder	10·0
Fish soluble, dried	5·0

Fig 11 Air blowers to increase oxygen supply

Fig 12 Trash fish or scraps minced with a meat grinder

Yeast powder	10·0
Vitamins	1·0
L-Lysine	0·1
D.L. Methionine	0·1
Binder	0·2
Anti-oxidant	0·2

Its chemical composition is as follows:

	%
Crude protein	45·43
Crude ash	14·66
Carbohydrates	21·34
Crude fibre	3·01
Water	9·35

The above artificial feed was compared with minced fish in a 90-day feeding experiment. At temperatures ranging from 9·6 to 18·6°C, the conversion rate of the artificial feed was found to be 2·45 as compared to 13·49 for trash fish. The experiment was performed on young eel of about 45 g in initial weight.[4]

In application, the artificial feed in powder form is well mixed with 5% to 10% of fish liver oil and 10% of water to form a stiff paste, which is then placed in the feed basket.

The quantity of feed given per day is 5% to 15% of the total weight of the eel in the pond in the case of minced trash fish and 1·0% to 3·5% in the case of artificial feed. The quantity is limited to what the eel can consume in 20 minutes.[3]

Because of its comparatively high cost, no eel farmers in Taiwan use the artificial feeds exclusively. Usually they keep artificial feeds in stock and use them when the supply of trash fish is low and its cost high.

8. HARVESTING AND MARKETING

From June to September, when some of the eels have reached marketable size, they are harvested daily or once every few days at feeding time by a net placed below the feeding platform and taken to the local market. This will reduce the population in the pond as well as provide a revolving fund to the eel farmer.

In addition, the eel ponds are drained at least once a year and all the eels taken out. This is for three purposes:

1. To remove all the eels of marketable size for export and for local

consumption. Eels of the size 5 to 8 per kilogramme are preferred by the export market and those of 3 or 4 per kilogramme are preferred by the local market.

2. To clean and disinfect the ponds and expose the bottoms to the sun.

3. To segregate the eels as to size and restock them in different ponds. This is important for uniform growth.

After the eels are removed from the ponds, those to be sold or exported are segregated by size and impounded in clean water for a few days to empty their guts and condition them for transportation (Fig 13).

Fig 13 Eels for market are kept in clean water for few days before transportation

Eels are sold either in the domestic market or export market. In inland transportation over short distances, the eels are first chilled in ice water to put them in a state of suspended animation and then put in bamboo baskets over which ice water trickles. For export to Japan, the previously chilled eels are put into a plastic bag with a small quantity of water and filled with oxygen. Two such bags are encased in a strong carton for shipment by air.

Elvers are shipped either in oxygen-filled plastic bags or in shallow plastic boxes with some ice and water. They are also impounded in clean water and chilled before shipment.

9. THE SEED EEL DILEMMA

The supply of seed eel for eel farming is the most perplexing problem faced by the eel farmers of Taiwan today. The average yearly catch of glass eel from the estuarine waters of Taiwan is about 15 mt, which are caught for stocking the 300 ha of eel ponds in the past. With the rapid increase of eel ponds, a serious shortage of seed eel has developed since 1971. The price of locally produced seed eel has gone up tremendously, and there is a rush to buy the less costly imported seed eels. In 1972, the price of each locally produced glass eel once went up to NT$22·00 (US$0·58) as compared to about NT$2·00 for an imported glass eel. In 1973, the Government lifted the ban on importation of glass eels that are not of the species *Anguilla japonica*, and a large quantity (estimated at 20 mt) has been imported since February. The price of each imported eel dropped to about NT$0·70 (US$0·02). Facing such strong competition, the price of locally produced glass eel was forced down to NT$12·00 (US$0·32) each.

Although the Japanese eel, the culture techniques of which have long been mastered by the local eel farmers, remains the favourite, the disparity in cost has influenced many eel farmers to buy the imported eel. They figure that even if the imported seed eel suffer a high mortality of 80%, they would still save money.

The short supply of seed eel is not a problem peculiar to Taiwan. Japan, the leading eel farming country and the largest eel consumer of the world, faces the same problem. The fish culturists of Japan are performing extensive experiments to utilize the imported seed eel of various origin for stocking their ponds.

Although some elvers from Australia and New Zealand have been imported into Japan for trial, the elvers of the European eel, *A. anguilla*, are still looked upon as the most promising and the largest source of supply. The European eel has a very large distribution in the North Atlantic and Mediterranean, extending from the coast of Denmark down to the north coast of Africa.[5]

The imported glass eel, of the species *Anguilla anguilla*, cannot tolerate a high temperature and is liable to many diseases and parasitism. The writer has visited many eel farmers who kept imported seed eel in their ponds and found that, in spite of the different culture methods used and intense care, such as (1) use of water of temperatures not exceeding 25°C, (2) the application of pesticides, antibiotics and sulpha drugs, (3) more frequent feeding and (4) larger flow of running water, most of them suffered losses. Others claimed success of varying degree, ie a survival rate of 70 to 90% up to the time of marketable size.

It is found that the techniques successfully used by the eel farmers vary in one way or another. Among eel farmers who use identical technique, some succeeded and others failed. One eel farmer who claimed success in 1971, failed utterly in 1972, employing the same techniques. This leads some eel farmers to believe that they are not dealing with eels of the same species or origin.

The imported seed eel are purchased from Japanese exporters. The buyers in Taiwan have no definite knowledge of their origin. France, Italy, Spain, Morocco, England, the Philippines, Indonesia and Sabah have been cited as the countries of their origin. Morphologically, it is well nigh impossible to distinguish the seed eel of *A. anguilla* from that of *A. japonica*. As a matter of fact, the *A. anguilla* is distinguished from the *A. japonica* by the application of the pesticide Ciodrin, the 1 ppm solution of which will kill the *A. anguilla* elver in 20 minutes leaving the *A. japonica* unharmed.

It seems necessary, therefore, for the Taiwan eel farmers to import seed eels directly from the countries of origin. A project should also be initiated to experiment on the rearing of eels of different origins and determine which is the more desirable.

For the purpose of minimizing disease and parasite infestation of the imported seed eel, the eel farmers of Taiwan dose their eels with different preparations of sulfa drugs, antibiotics, pesticides and growth promoting substances. This practice of trial and error is not only uneconomical but will undoubtedly impair the health of the eels. In this respect, scientific investigation should be instigated to look into the cause of each disease and parasitism and institute treatments and preventive measures.

LITERATURE CITED

1. Kuo, Ho, Chi-Tsai Huang and Ying Yao Su: Survey of Elver Production in the 1971–72 Season. *China Fisheries Monthly*, No. 242, Feb. 1973. (in Chinese.)
2. Chen, T. P.: The Fabulous Eel Industry. *Industry of Free China*, April 1971, pp. 41–42.
3. Kuo, Ho: Eel Farming. *Extension Guide Book No. 72A of Taiwan Provincial Department of Agriculture and Forestry.* (in Chinese.)
4. Lai, Y. H. and Y. Y. Su: Experiment on the Efficiency of Artificial Feed for Eel. *China Fisheries Monthly*, No. 177, 179, 182, 188. (in Chinese.)
5. Deelder, C. L.: Synopsis of Biological Data on the Eel *Anguilla anguilla. FAO Fisheries Synopsis No. 80.*

III. Polyculture of Chinese Carps in Freshwater Ponds

1. INTRODUCTION

Polyculture of carps is the backbone of fish culture in China. It began thousands of years ago and has been improved with the passing of time. William E. Hoffmann was perhaps the first westerner to describe it in detail.[1] It is particularly common in the delta region of the Pearl River and the Yangtse River, where natural supply of the fry is plentiful and the general conditions favourable.

In Taiwan, the total acreage of freshwater fish ponds in 1972 was 10,275 ha, most of which were used for the polyculture of Chinese carps together with other suitable species. In order of importance, the species of Chinese carps planted are grass carp, *Ctenopharyngodon idellus* (Fig 14), silver carp, *Hypothalmichthys molitrix,* bighead, *Aristichthys nobilis,* mud carp, *Cirrhina molitorella,* and snail carp, *Mylopharyngodon piceus*. Other fishes often reared with the Chinese carps are mullet, common carp (Fig 15), Crucian carp, tilapia, etc.

Polyculture of Chinese carps is believed to be capable of achieving the highest productivity in a given unit area, because, first, these fish occupy different strata of water: the mud carp and the snail carp generally stay at the bottom, the silver carp and the bighead usually at the top, and the grass carp roam all strata of the water. Secondly, they have different feeding habits: the grass carp feeds on grass and other plant materials, silver carp on phytoplankton, bighead on zooplankton, mud carp on detritus on the bottom, and snail carp (also known as black carp) on live food, principally molluscs and crustaceans on the bottom.[2]

When these Chinese carps are reared in the same pond, therefore, there is a saving in feeds as well as space. The ratio or number of each species planted in the pond varies with the condition of the pond and management practice and will be discussed later.

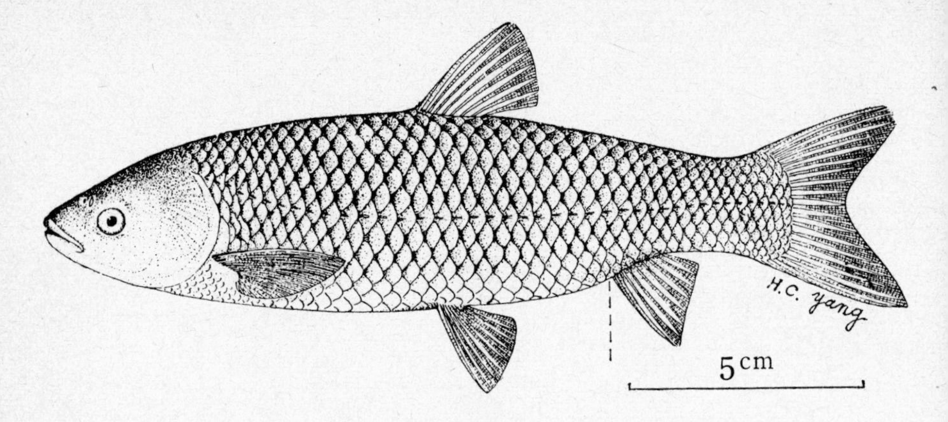

Fig 14 Grass carp *Ctenopharyngodon idellus*

2. SUPPLY OF FRY OR SEED

The five species of Chinese carps have their origin in the rivers of China, mainly the Yangtse River and the West River (also known as Pearl River). The spawning grounds of the mud carp, however, do not extend to the Yangtse area, and are limited to the rivers of South China. Fertilized eggs are carried down by the streams, hatch on the way, and are collected as fry.[3]

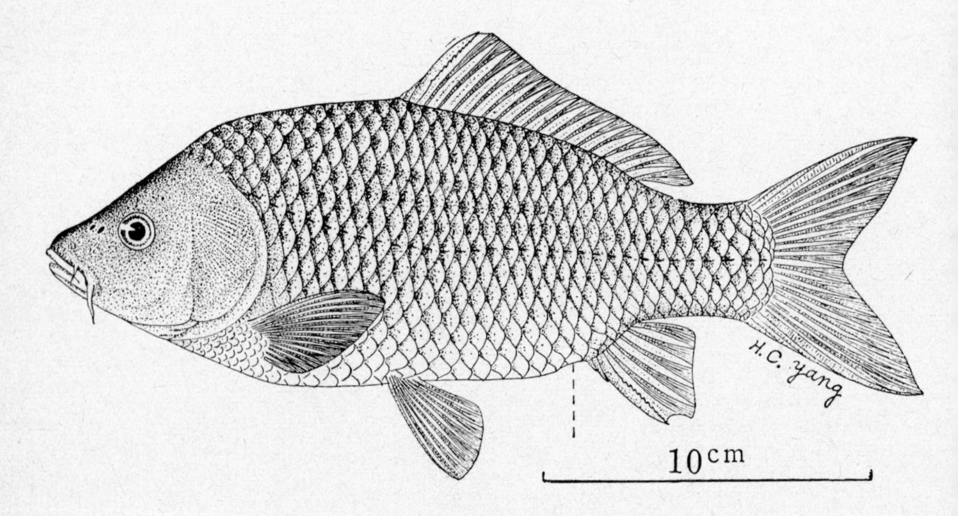

Fig 15 Common carp *Cyprinus carpio*

As the Chinese carps do not spawn naturally in confinement, Taiwan, until 1964, had to depend on fish seeds imported from Hong Kong, and to a small degree from Japan, where the grass carp and silver carp, introduced from China, had started to spawn in the Tone River. The number of fry imported was from 30 to 40 million each year.

In 1962, with the success in induced spawning of the grass carp and silver carp by the Tainan Fish Culture Station, the Taiwan Fisheries Research Institute began to produce fingerlings of Chinese carps, and this was soon followed by a large number of small commercial hatcheries. So, beginning from 1964, Taiwan was not only self-sufficient in the supply of seeds of all the species of Chinese carps, but was also able to export them to Southeast Asian countries as well as small quantities to the United States.

3. PRODUCTION OF FRY BY INDUCED SPAWNING

The season for the production of the fry of the Chinese carps in Taiwan begins in March and continues to July. The commercial producers have now completely mastered the technique of artificial propagation.[4]

The first and most important step in the artificial propagation of the Chinese carps is the culture of spawners. From the stock of Chinese carps in ordinary commercial ponds, spawners are selected for use in induced spawning (Fig 16).

Grass carp intended to serve as spawners should be given plenty of feeds such as tender grass, vegetables, aquatic plant, peanut meal, soybean meal and rice bran. Over-feeding, however, will cause the fish to grow fat with adipose tissues accumulated on the gonads, thus hindering egg and sperm development.

For silver carp, bighead and mud carp, peanut meal and soybean meal are good feeds, and the brood ponds should be fertilized with inorganic fertilizers and barn manures.

Under the climatic conditions in Taiwan, the females of silver carp become fully mature in two or three years (2 kg or more), bighead three to four years (5 kg or more), grass carp four to five years (3 kg or more) and the mud carp three to four years (1 kg or more). The males mature one year earlier than the females.

There is practically no commercial production of the fry of the snail carp, but small numbers are produced by the Taiwan Fisheries Bureau's hatchery on Coral Lake. Snail carp is not popular in Taiwan and the demand for its fry is therefore small.

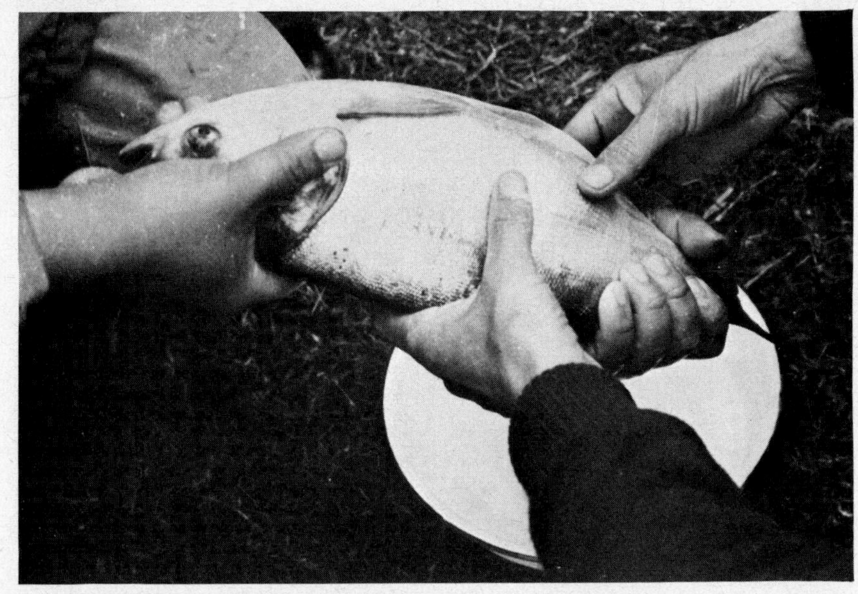

Fig 16 Selection of silver carp spawner—
eggs being stripped

Fig 17 Mixing sperm and eggs

Spawners of Chinese carps are held in small mud ponds of about 1,000 square metres in area with a depth of about 1 m. Several holding ponds should be provided, so that the spawners in each pond could be netted and used in rotation. This will minimize injury to the fish. In most cases, the spawners will die from shock and injury after stripping, but with tender care the same spawner could be used repeatedly for many consecutive years.

The next step is to ascertain the ripeness of the spawners. For the males of all Chinese carps, a few drops of milt by application of light pressure to the abdomen is indication of ripeness. Ripeness is also indicated by secondary sexual characteristics. Serrated ridges, rough to the touch, occur on the inner surfaces of the pectoral fins of the males of the mature silver carp and bighead. Similar characteristics are also found in the mature males of the grass carp but less prominently. Gravid femals can be recognized by the soft distended belly caused by full growth of the ripe ovary, and the swollen anus, pinkish in front.

Spawning is induced by the injection of pituitary of cyprinids, generally the common carp, with Synahorin, Gonagen-forte or Puberogen as a booster. The dosage and procedure vary a great deal according to the condition of the spawner and the experience of the operator. In case a spawner is fully mature and in good condition, a small dose would be effective; otherwise a heavier dose would be needed. In principle, however, the weight of the donor of the pituitary (or several donor fish) should be equal to that of the recipient fish. The dosage of Synahorin is generally 12 to 14 rabbit units for each kilogramme of the spawner. The pituitary and Synahorin mixture is divided into two portions for injections at interval of six hours.

Usually no injection of hormone is given to the male fish unless milting is found inadequate, in which case an injection of similar dosage given to the female is administered at the time of the second injection.

Differing from the other Chinese carps, the mud carp is given only one injection, and the males and females are placed together in one pond, where natural spawning and fertilization will take place after about eight hours.

After the injections, when the female is ready to spawn, it is stripped by pressure on the abdomen (Fig 16). The eggs, in an enamel pan, are fertilized with sperm stripped from a ripe male by mixing gently with a feather (Fig 17). The fertilized eggs hatch in about 30 hours at a temperature of 21–24°C, and about 19 hours at 28–30°C. Lin's hatching net (a conical net with water coming up from a vent at the bottom to keep the eggs rolling) developed by the late Mou-chun Lin (Fig 18) is used for hatching.

The hatchlings are about 6 mm in length (4·5 mm for the mud carp), transparent, with large yolk sacs. They are left in the hatching net with running water for about two days (Fig 19), after which they are transferred to a rearing trough of fine-mesh netting suspended in a pond and fed with egg yolk, soybean milk, skim milk powder, wheat flour, etc. After three or four days, they are transferred to an earth pond, from which they will be sold to fish fry dealers.

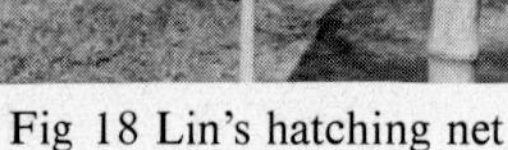

Fig 18 Lin's hatching net

Fig 19 Hatchlings kept in nets for two days

4. REARING OF THE FRY

Most carp fry dealers are located in the Tainan and Changhua areas. They buy the hatchlings from the hatcheries and sell them to fish farmers or fish fry exporters. They handle not only the Chinese carps, but also the fry of many other species of freshwater pond fish, including snake-heads, walking catfish, Crucian carp, leather carp, Tilapia, etc. They generally have a number of nursery ponds for holding the fry.

Before stocking with fish fry, the nursery ponds are drained, dried under the sun, fertilized, and rid of undesirable organisms (by application of lime or tea-seed meal). After the ponds are refilled with water, rotifers, daphnia and other natural foods will develop in quantity to serve as feeds for the fry in seven to ten days, depending on the weather.

The nursery ponds are stocked at the rate of not more than 300,000 per tenth of a hectare. But sometimes 500,000 are planted in one tenth of a hectare for purpose of stunting the fish. They grow to 2–3 cm in length after about ten days, when they should be stocked in ponds at the rate of 70,000–80,000 per tenth of a hectare. The stocking rate should be further reduced when the fish attain length of 4–5 cm. The fry may be used as stocking material when they are 5–10 cm in length.

About three days after the nursery ponds are stocked, the natural food should have been all consumed, and supplementary feeds such as soybean milk, egg yolk, milk powder and liver extract should be given. A week later, these feeds are replaced by ground peanut meal, soybean meal, wheat flour or rice bran in the quantity of 4–10% of the total weight of the fry.

The most serious pests in the nursery ponds are the predacious diving beetle, *Hydaticus* sp., and tadpoles. The former may be eliminated by the application of 0·5 ppm Dipterex, 0·3 ppm Sumithion or 0·25 ppm Lebaycid. To reduce the number of tadpoles, the ponds may be fenced to prevent the entry of frogs, and any frog spawns found in the ponds should be removed.

5. TRANSPORTATION OF THE FRY

Less injury occurs to the fry during transportation when they are about 20 days old or when they are over 30 days old. Before they are shipped, the fry must first be conditioned so as to reduce mortality during the journey. Conditioning is done by towing a fine-meshed seine across the pond and holding the fry thus collected in the seine at high concentration for about five minutes before releasing them. After an interval of about 24 hours, the fry are again collected and placed in a cage net suspended in the pond (Fig 20) for about six hours. For transportation over long distance, the fry should be further held in clear water for three hours or longer. No feeds are given during conditioning. Conditioning will accustom the fish to crowding and also cause them to empty their guts of food.

Plastic (PVC) bags are now the universal container used for transporting the fry. They are 40 cm × 30 cm × 120 cm high. Each bag holds 500 fry of 7 cm, 1,000 fry of 5 cm or 8,000–10,000 fry of 2·5 cm in 10 litres of water for a short haul of not more than ten hours. In hot weather, 1 kilogramme of ice is added to the water.

6. REARING FOR MARKET

Two types of culture are practised in rearing the Chinese carps for market, (1) extensive culture or 'rough release' in reservoirs or large water surfaces and (2) intensive culture in ponds. With the former, growth of the fish depends on natural feeds produced; fertilizers and supplementary feeds are only occasionally given. The turnover of water in irrigation reservoirs affects the yield and makes it impractical to fertilize the reservoir ponds.

The yield is therefore comparatively low (not exceeding 1,000 kg/ha), except at places where there is considerable influx of sewage, in which case the yield may be much higher (about 2,000 kg/ha). The following procedure describes the culture in ponds under intensive management and has given exceedingly high yields.

6.1 Selection of pond site

The best ponds for Chinese carps are located in areas with the following qualifications:

(1) Free from inundation;
(2) Long exposure to sunshine;
(3) Adequate supply of water – either surface or underground;
(4) Supply of good quality water (free from chemical pollution, low iron content, salinity lower than 1‰, pH 7–8·5);
(5) Loamy soil with little seepage;
(6) Close to transportation facilities.

6.2 The fish pond

The ideal size of the pond is from 0·5 to 2·0 hectares. A pond smaller than this will not provide the space required for rapid growth, and the smaller body of water will be subject to a rapid change of temperature and other water qualities. On the other hand, a large pond is more difficult to manage, eg in harvesting and pest control. It is preferably rectangular in shape.

Sluice gates are placed both at the water inlet and outlet, which are at opposite sides of the pond. Both should be screened to prevent the entrance of undesirable organisms as well as the escape of fish. The pond bottom should be flat with a gentle slope of about 0·5% to effect complete drainage of water whenever required. A trench should be provided near the outlet to collect the fish when the pond is drained. It is desirable to maintain the depth of water at 1·5–2·0 m.

6.3 Stocking

Ponds which have been in use for years are drained after the winter harvest, the debris and decomposed organic matters removed, the dikes repaired, and the bottom dried under the sun. To each hectare of pond, 1,000 kg of lime is then added to eradicate pests and predators. This winter treatment of the pond is important to maintaining the fish yield.

The ponds are stocked generally in March and April. As propagation of the Chinese carps does not begin until late March, the stocking materials held at the time are over-wintered fingerlings of 7 to 14 cm in length.

Both stocking rate and stocking ratio, as well as the species to be planted vary a great deal according to locality, the fertility of the pond, the supply of natural and supplementary feeds, and the manner of operation. Generally speaking, silver carp, bighead, common carp, Crucian carp and tilapia are suitable species for eutrophic ponds (ponds rich in nutritive matters), particularly when duck-cum-fish farming (Fig 21) or hog-cum-fish farming is practised. In oligotrophic (poor in nutritive matters) ponds, especially when grass and weeds grow in abundance, the grass carp should be the dominant species to be stocked. In ponds with an abundance of extraneous fish, small number of predatory fish such as sea perch (Fig 22) and snakehead may be introduced. Because of temperature difference, tilapia are generally not stocked in ponds of northern part of Taiwan. On the other hand, the euryhaline milkfish are frequently stocked in the freshwater ponds of southern Taiwan.

It is difficult to give a formula of stocking rate and stocking ratio of freshwater ponds in Taiwan. But it is possible to give two examples as follows:

Table 3 Stocking rate of freshwater pond (per hectare)

Species	Size	No. of fry	Time of stocking
Example A (Central Taiwan)			
Silver carp	7–12 cm	800	March–April
Bighead	7–12 cm	100	,, ,,
Grass carp	7–12 cm	50	,, ,,
Mud carp	5 cm	1,000	,, ,,
Mullet	5 cm	2,000	,, ,,
Common carp	2·5 cm	1,000	,, ,,
Example B (Southern Taiwan)			
Silver carp	10–13 cm	1,000	February–April
Bighead	10–13 cm	400	March
Grass carp	12–15 cm	200	March
Mud carp	7–10 cm	1,500	March
Common carp	3–4 cm	2,000	March
Crucian carp	3 cm	2,000	February
Mullet	5 cm	2,000	March
Walking catfish	5 cm	500	May
Snakehead	10 cm	500	June

6.4 Feeds and fertilizers

The traditional feeds such as rice bran, soybean and peanut meal are given to supplement the natural foods produced in the ponds. Feeding is once daily. The amount of feeds given is about 2% of the total weight of the fish. As the optimum temperature for feed utilization is 20–28°C, more feeds should be given within this temperature range. As the winter temperature drops below 15°C, feeding activity of the fish diminishes and comes to a stop at about 5°C. So, feeding should be regulated according to temperature changes. As a rule, no feeds are given during the cold days in winter.

Fig 20 Cage net for conditioning fry

Fig 21 Duck-cum-fish ponds

The grass carp are mainly plant eaters. Before they are old enough to eat the tougher and larger aquatic or land plants, the grass carp fingerlings are fed with duckweed, *Lemna*. Some farmers in the Tainan area grow duckweed and sell it to fish farmers. The older and larger grass carp, if stocked in sufficient number, will soon clean the pond of vegetation, and cuttings of land plants should be supplied as feed.

As both the silver carp and bighead are plankton feeders, the ponds should be fertilized to promote the growth of phytoplankton. The use of organic fertilizers such as barnyard manure, night soil, rice bran and soybean or peanut meal is still universal in Taiwan, in spite of efforts to promote the use of inorganic fertilizers. At present inorganic fertilizers are only popularly used in central and northern Taiwan, particularly in Taoyuan and Hsinchu Hsien, where the supply of organic fertilizer is inadequate, inconvenient or too expensive.

Phosphorous is the most common limiting factor of the production of natural foods in fish ponds. Application of phosphorous has been found to increase the yield of fish tremendously in otherwise barren ponds, particularly where silver carp and bighead are the dominant species in the ponds. Application of slaked lime will also increase production in acid ponds.

Chinese fish farmers are experts in conditioning the water in fish ponds.[5] They judge the quality of the water by its colour. Green water means abundance of phytoplankton, which is food for the silver carp as well as for the zooplankton, which is food for the bighead. It also means abundance of dissolved oxygen produced by photosynthesis of the phytoplankton.

Experiments on the effectiveness of the addition of fly ash as a source of minor elements and of acetic acid as a source of carbon to the fish ponds have been carried out, but no significant results have been reported. The addition of Zeolite, a commercial preparation containing silica (SiO_2), to the pond water to promote the growth of diatoms has been tried with some success.

As mentioned, the depth of water should be maintained at 1·5 to 2·0 m. At this depth or more, the lack of adequate light penetration will discourage the growth of higher aquatic plants, and the application of phosphorous fertilizers at intervals of not over five days will maintain the plankton in good growth.

Finally, it should be stated that when organic fertilizers are applied, they also serve as supplemental feed either in part or totally.

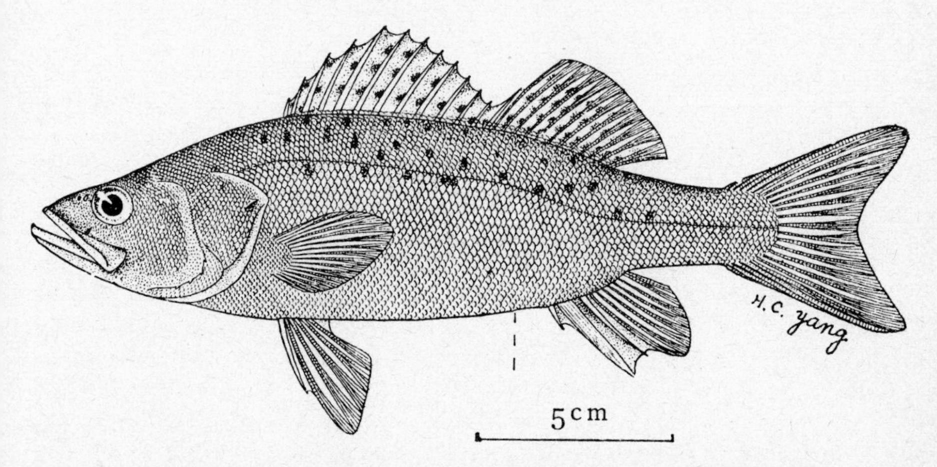

Fig 22 A few sea perch *Lateolabrax japonicus* are put into ponds

6.5 Diseases and parasites

The most prevalent disease in polyculture of Chinese carps is the inflammation of the intestine of grass carp. The wall of the intestine becomes swollen and is marked with pink blotches. A larger quantity of thick milky fluid can be easily pressed out through the seriously inflamed anus. The fish die in a few days. In a study,[6] it was found that the bacteria *Aeromonas punctata* was associated with the disease and that Chloramphenicol is effective in inhibiting the growth of this pathogen. The common treatment, however, is to incorporate sulpha drugs (such as Sulphaguanidine, 100 mg per kg of fish, or Chloramphenicol, 10 mg per kg of fish) in the feeds given to the fish. The drug is mixed with sweet potato or wheat flour paste which is applied to the feed grass. As a prevention, this is applied once every one or two months.

Gill rot is also common with the grass carp. It is a protozoan disease. Effective treatment is 8 ppm copper sulphate bath for 20 minutes.

The most common parasites of the Chinese carps and other pond fishes are the anchor worm, *Lernaea* sp. and the fish lice, *Argulus* sp. Their favourite hosts are the silver carp and bighead which have smaller scales they can penetrate easily. Their infestation weakens the fish, making them susceptible to disease, and is often fatal.

The larvae of the *Lernaea* can be killed with 0·5 ppm Dipterex, while 0·5 ppm BHC is effective in treating *Argulus* infestation.[7]

Fungus infection occurs usually in the winter and usually disappears when warm weather sets in. The common treatment is 0·1–0·2 ppm Malachite green or 1–2 ppm Methylene blue.

6.6 Harvesting

In polyculture of Chinese carps, the time-honoured practice was to drain the ponds at the beginning of winter (when the fish cease to grow) and harvest all the fish at once. This practice is now no longer followed, especially with the inclusion of tilapia in the pond. The common practice now is to net the fish several times in a year so as to avoid over-crowding as well as to meet special market demands during certain festivals. The tilapia (in southern Taiwan) are netted from time to time after the month of June. The silver carp that exceed 500 gm in weight are taken out to meet the demand of the Moon Festival (in September). The other fish are not harvested until around December and January for sale during the New Year season. (The Lunar New Year observed by the Chinese usually occurs in January or February.)

For climatic reason and for the reason that tilapia are stocked with the carps, the ponds in southern Taiwan are more productive than those in northern Taiwan. Four to five tons of fish per hectare are the usual production in Kaohsiung and Tainan, with a high of 10 tons/ha.

LITERATURE CITED

1. Hoffmann, William E.: Preliminary Notes on the Freshwater Fish Industry of South China, Especially Kwangtung Province. *Lingnan University Science Bulletin* No. 5, January 1934.
2. Chen, T. P.: A preliminary study on association of species in Kwangtung fish ponds. *Lingnan Science Journal,* Vol. 13, No. 2, April 1934.
3. Chen, C. S. and S. Y. Lin: The Fish Fry Industry of China. *Bulletin of Chekiang Provincial Fisheries Experiment Station,* Vol. 1, No. 4, November 1935.
4. Lin, S. Y.: Induced Spawning of Chinese Carps by Pituitary Injection in Taiwan. *JCRR Fisheries Series* No. 5, 1965.
5. Lin, S. Y.: Fish Pond Fertilization and the Principle of Water Conditioning. *China Fisheries Monthly,* No. 209, May 1970. (In Chinese.)
6. Wu, Wilson Sheng-yu: A Disease of the Grass Carp (*Ctenopharyngodon idellus*) and Its Chemotherapeutical Control. *JCRR Fisheries Series* No. 11, 1971.
7. Li, Yen-pin and Shiu-nan Chen: Some Parasites Found in Pond Fishes of Taiwan. *JCRR Fisheries Series* No. 12, 1972.

IV. Culture of the Grey Mullet

1. INTRODUCTION

There is virtually no monoculture of the grey mullet, *Mugil cephalus,* in Taiwan. They are nearly always reared in the same pond with Chinese carps and other freshwater fishes, and no special management practice is required aside from that for the polyculture of Chinese carps. As the artificial propagation of the grey mullet is more successful in Taiwan than in other areas of the world, greater emphasis will be placed in this chapter on the artificial production of mullet fingerlings than on pond culture.

Eleven species of mullet are found along the coasts of Taiwan, but the grey mullet, *Mugil cephalus,* is the species commonly caught in the sea and reared in ponds (Fig 23). It has a wide distribution, occurring in all tropical and subtropical waters around the world. In Taiwan waters, it is most abundant during January and February, when the mullet schools follow the coast line of Taiwan moving from north to south in their spawning migration. The mullet schools are most heavily concentrated in the waters off Kaohsiung and Pingtung. As these fish are all mature, they are caught mainly for the highly priced roes of the females.

2. ARTIFICIAL PROPAGATION

2.1 Historical survey

The experimental work on the artificial propagation of the mullet begun in 1964 has been carried out first in Sanwei of Kaohsiung County by a research team organized by the Taiwan Fisheries Bureau and later in Tungkang by the Tungkang Marine Laboratory of the Taiwan Fisheries Research Institute. These strategic locations enabled the research workers to obtain live ripe mullet quite easily from the fishermen.

In 1963 the Taiwan Fisheries Bureau organized a team of workers from the Taiwan Fisheries Research Institute, National Taiwan

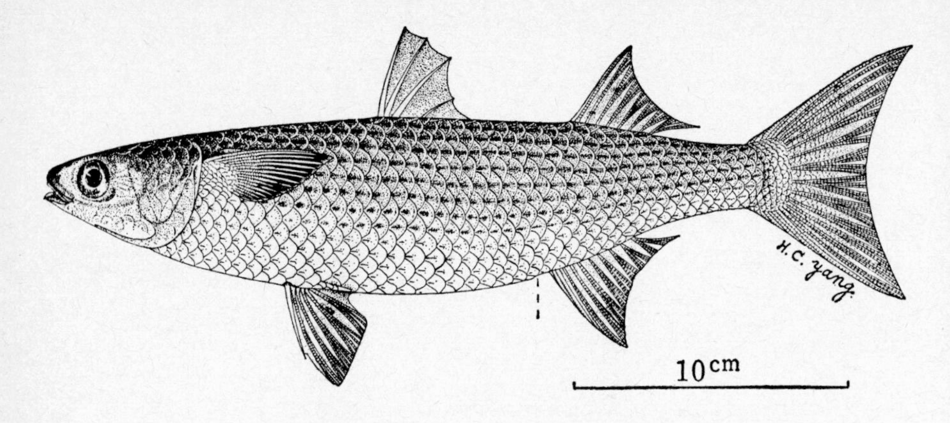

Fig 23 Grey mullet *Mugil cephalus*

Fig 24 Research workers select healthy ripe mullet caught by purse seiners

University and Taiwan Fisheries Bureau to carry out experiments on the artificial propagation of the mullet.[1] The experiments were continued each winter during the mullet migration season except in 1968 when no funds were made available for this work. During 1963 and 1964, the experimental work was mostly on the selection, transportation and impoundment of the spawners and hormone treatment. In 1965–1967, experiments in the rearing of the larvae were started. In 1969, two fingerlings were reared up to the 31st day; they measured 1·0 and 1·1 cm, respectively, and were covered with scales. In 1970,431 fingerlings of size suitable for stocking were produced. The numbers of mullet fingerlings of stocking size produced in the following years were 1971, 7,786; 1972, 23,695; 1973, 21,688; 1974, 6,050; and 1975, 13,916, respectively.

2.2 The spawners

The spawners are obtained from amongst the mullet that migrate each winter to the waters off the coast of Pingtung County to spawn.[2] The research workers go out to sea on an outboard powered raft to select and procure healthy ripe fish (Fig 24) from purse-seiners. The breeders so procured are placed in dark coloured plastic bags filled with sea water and oxygen and brought back to the stock tank of the Laboratory. Most of the mullet caught belong to the IV-year class, measuring 32–50 cm in body length and weighing 1·0–2·1 kg each.

The stock tank is an indoor concrete tank measuring 5 m × 7 m × 1·5 m (Fig 25). In the tank, the males and females are separated by a nylon net. Fresh sea water is continuously introduced into the tank, and adequate aeration is provided.

2.3 Induced spawning

The materials injected to induce spawning are (1) pituitary gland obtained from mature mullet (either male or female) (Fig 26), which may be preserved in acetones at 5°C and (2) Synahorin, which is a mixture of chorionic gonadotrophin and mammalian hypophyseal extract.

The best result is obtained by giving the female fish the first injection within an hour after its introduction into the stock tank and the second injection within the next 24 hours (Fig 27). A third or a fourth injection is given if there is no response after the second or third injection. According to Liao,[1] successful ovulation is often obtained by injecting 2·5 to 6·0 pituitary glands combined with 10 to 60 rabbit units of

Fig 25 Indoor concrete tanks for holding spawners and rearing larvae

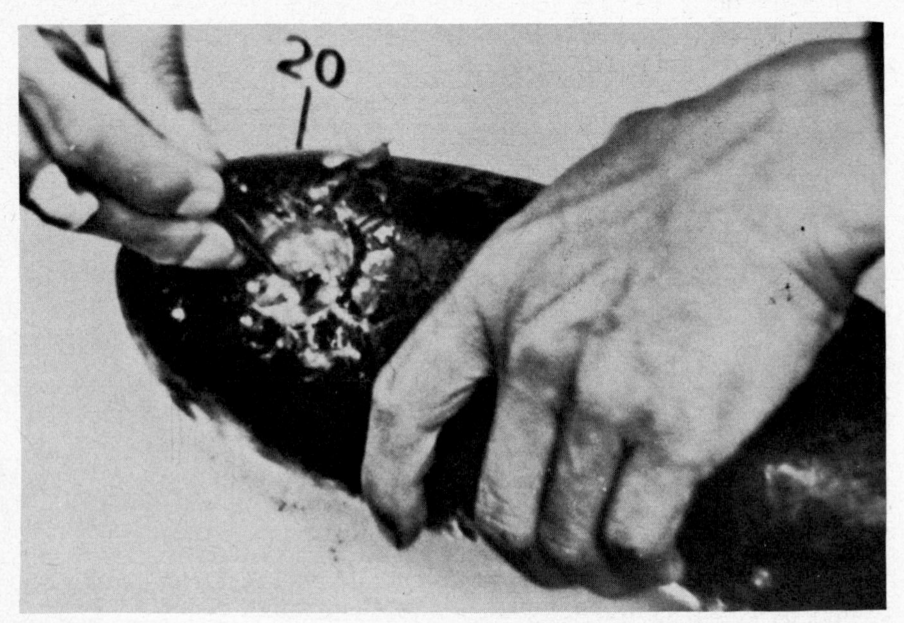

Fig 26 Removing pituitary gland from donor fish

Synahorin and 0 to 300 mg of vitamin E. Intramuscular injection on the dorsal part of the fish is commonly practised.

For the male mullet, hormone treatment is only needed for fish caught toward the end of the spawning season. The majority of the male mullet caught are fully ripe and ready to yield milt without any hormone treatment.

Fully ripe and healthy fish respond readily to hormone treatment. The belly of the female becomes greatly extended, and the eggs come out freely through the genital pore when light pressure is applied to the belly, sometimes even without pressing. To examine the condition of the eggs, one usually sucks some eggs from the genital pore with a pipette and examines them under a microscope. If the eggs are transparent and completely round with one oil globule, they are ready to be fertilized.

One female of 1·5 kg usually yields 1 to 1·5 million eggs.

The female fish is stripped by one person and the eggs are collected in a plastic basin (Fig 28). Another person strips the male and lets the milt fall onto the egg mass. A third person mixes the eggs and milt gently with a feather (Fig 29). The fertilized eggs are washed in several changes of sea water to remove blood and other foreign matters. Then they are put into water in plastic tanks with aeration to hatch. Either the 'dry method' or 'wet method' of artificial fertilization can be used. The only difference is that fertilization could be done any time within an hour using the 'dry method', but it must be done within five minutes in case the 'wet method' is used.

The fertilized egg is round, transparent, non-adhesive, with a large yellowish oil globule measuring about 0·38 mm in diameter. The egg is 0·93 to 0·95 mm in diameter. The fertilized eggs stay afloat near the surface of the water under slight aeration, but some may settle down slowly in still water. Dead eggs sink to the bottom.

2.4 Hatching

Plastic tanks of 0·5 and 1·0 ton capacity and large indoor concrete tanks (5 m × 7 m × 1·5 m) are used for hatching the fertilized eggs (Fig 30). The water temperature is maintained at 20–24°C, and the water quality is maintained by continuous change of water and continuous aeration. High dissolved oxygen content, minimum variation in water temperature, cleanliness and gentle movement of the water are essential factors of success. After 16–30 hours of incubation, the eggs will have well-developed embryos with obvious black pigments.

The rate of cleavage and embryonic development varies with the

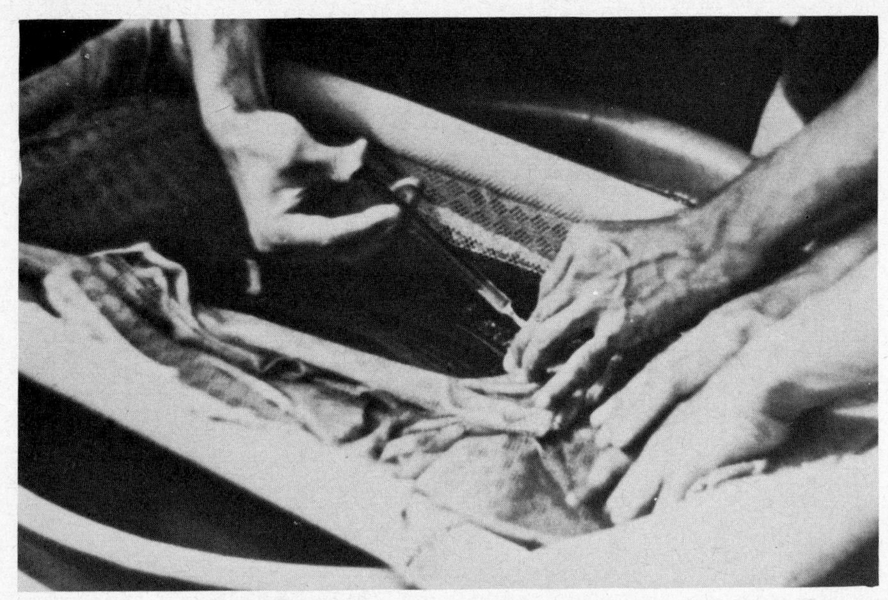

Fig 27 Injecting pituitary and Synahorin into mullet

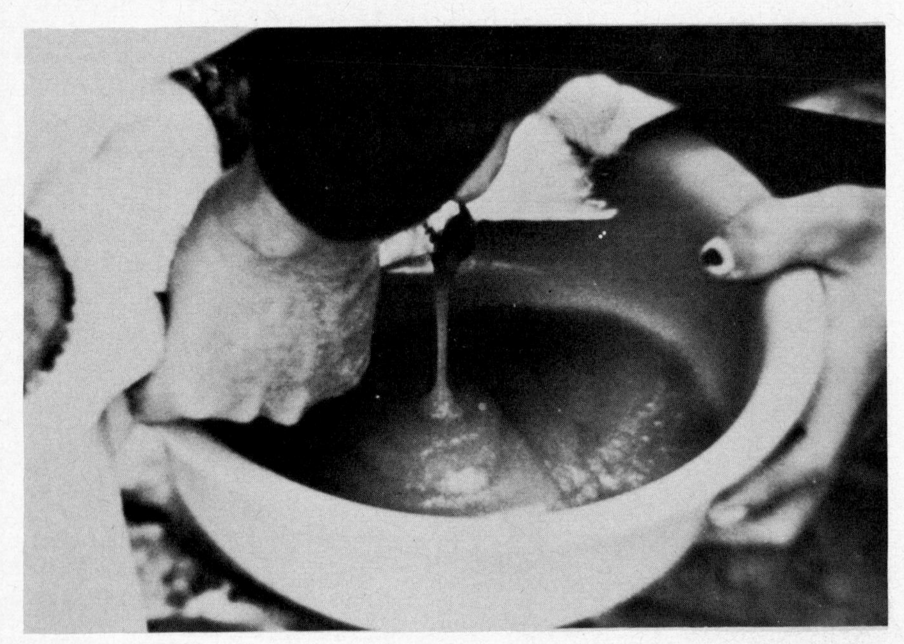

Fig 28 Stripping female mullet

temperature of the water in which the eggs are incubated. The eggs will hatch in 34–38 hours at temperature of 23–24·5°C and 49–54 hours at 22·5–23·7°C, with salinity ranging from 30·1 to 33·8‰.

2.5 Rearing the larvae

This is the most difficult part of the whole operation. The newly hatched larvae are very small, ranging from 2·5 to 3·5 mm in size (Fig 31). They are transparent with complete fin fold and dark chromatophores scattered throughout the entire body. Their eyes are colourless, and the mouth and digestive tract are not well-developed. They show weak swimming activity with belly up and head down, sometimes with an up and down darting movement. Young larvae do not like strong light and tend to congregate in places of low light intensity. Older larvae swim in schools.

Supply of suitable feeds is the most important problem in rearing the mullet larvae. Different kinds of feeds are used according to the progressive development of the larvae as shown below:

Days after hatching	*Kind of feed*
3rd to 13th	Fertilized oyster eggs and trochophore larvae
5th to 18th	Rotifers (mostly *Brachionus plicatilis* collected from brackish-water fish ponds)
10th to 40th	Copepods (minute species or larval stages collected from brackish-water fish ponds)
16th to 40th	First *Artemia nauplii* and later the adults
29th to 44th	Cooked egg yolk, rice bran, wheat flour, etc.

It will be noticed that several kinds of feeds are sometimes given within the same period when one feed is to be replaced by another.

Thus far the mullet larvae are reared in indoor tanks. After the 40th day, they will generally grow to fingerlings of 1·5 to 2·0 cm in length and may be moved to outdoor ponds (Fig 32).

3. SUPPLY OF FINGERLINGS FROM NATURAL WATERS

Up to the present, practically all the mullet fingerlings used for stocking the commercial fish ponds are caught from natural waters, although the Tungkang Marine Laboratory has succeeded in the mass production of mullet fingerlings by induced spawning. The season for catching the mullet fingerlings is from October to March. Four kinds of fingerlings are

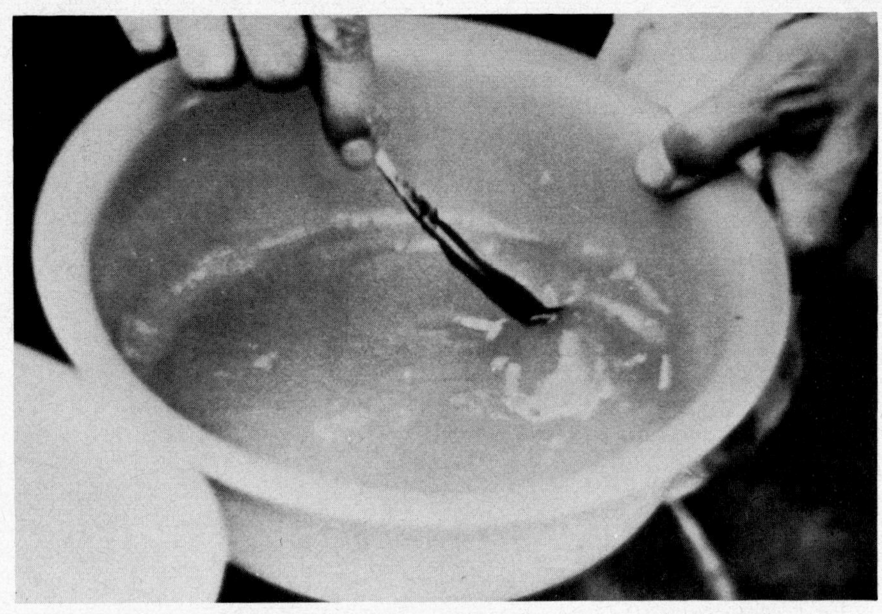

Fig 29 Mixing sperm and eggs with feather to ensure fertilization

Fig 30 Hatching tank with aeration equipment

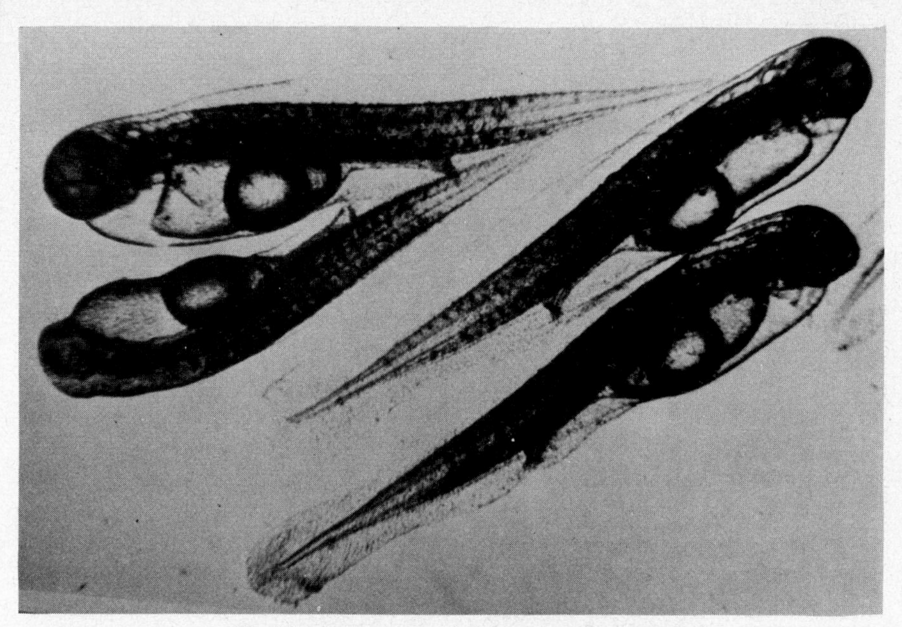

Fig 31 Mullet larvae 8–10 hours after hatching

Fig 32 Hatchery reared mullet 1 year old

differentiated, although they evidently belong to the same species. Those caught from October to December are locally called Ya Seng Tse, those in January are called Ta Chin Lin, those in February Yin Po Tse, and those in March Ching Tse Tse. The fish farmers prefer the Ta Chin Lin, which they claim to be faster growing than the others and give the best results.

Two kinds of gear are used to catch the mullet fingerlings, which generally congregate in estuarine waters. A small floating drag net is used in deeper water, and a small beach seine is used in shallow water.

The fingerlings caught are kept in fresh water for two weeks before they are stocked into the rearing pond. This will increase the survival rate. Mullet fingerlings are not sensitive to change of salinity but are very sensitive to sudden changes of water temperature. Care must be taken to see that any change of water is gradual.

4. POND CULTURE

As stated earlier, nearly all mullet are polycultured with Chinese carps, tilapia, etc. in freshwater ponds. According to the Taiwan Fisheries Year Book,[3] in 1973 there were 985 ha of ponds in which mullet as the dominant species is cultured with other fish, of which only 54 ha were brackish water ponds. The reasons for this are:

a. It is not desirable to culture mullet together with milkfish, the only brackish-water fish cultured in Taiwan, as it competes with the milkfish for food and would destroy the algal beds.

b. Most fish farmers believe that the mullet grow better in fresh water, although the Tungkang Marine Laboratory, in a recent experiment, demonstrated that the growth of the mullet was better in sea water and brackish water than in fresh-water ponds.[4]

4.1 Stocking practice

Mullet are stocked in polyculture ponds with grass carp, silver carp, big-head carp, common carp, tilapia, etc. The stocking practices vary a great deal, depending largely on the condition of the pond and the fish community in the pond.

The mullet is a benthic feeder with a stomach not unlike the gizzard of a chicken for grinding the detritus which the fish picks up from the bottom, so in a fertile pond with a bottom rich in organic matter more mullet may be stocked.

Then if more of the other species of fish are stocked, the number of mullet should be reduced. This is especially the case if a large number of common carp and mud carp, both benthic feeders, are placed in the pond. Generally, however, 1,000 to 2,000 per ha are stocked into ponds of polyculture, and 4,000–10,000/ha for ponds of monoculture.

4.2 Feeding and pond treatment

Since the grey mullet are reared in the same pond with carps and other fish, no special feeds are given to them and the ponds are treated in the same way as any polyculture ponds for Chinese carps.

LITERATURE CITED

1. Liao, I. C. Y. J. Lu, T. L. Huang and M. C. Lin: Experiments on Induced Breeding of the Grey Mullet, *Mugil cephalus* Linnaeus. *Symposium on Coastal Aquaculture, Indo-Pacific Fisheries Council,* 1970.
2. Liao, I-chiu *et al:* Preliminary Report on the Mass Propagation of Grey Mullet, *Mugil cephalus* Linnaeus. *Joint Commission on Rural Reconstruction Fisheries Series* No. 12, 1972.
3. Taiwan Fisheries Bureau: *Taiwan Fisheries Year Book,* 1973.
4. Liao, I-chiu: Unpublished data.

V. Tilapia Culture

1. INTRODUCTION OF TILAPIA INTO TAIWAN

The Java tilapia, *Tilapia mossambica* (Fig 33), was first introduced into Taiwan from Indonesia by the Japanese in 1944, when both Taiwan and Indonesia were under Japanese occupation. The fish were planted in the freshwater ponds in Pingtung, the southernmost (also warmest) county of the Island. But they did not spread to other areas and were not given much publicity. In 1946, Wu Chen-huei and Kuo Chi-hsin, two Taiwanese returning from Singapore, brought with them thirteen *Tilapia mossambica* to Kaohsiung. They placed them in their own ponds, and the tilapia multiplied and spread until they became a common food fish in all parts of the island.[1]

Following the introduction of *T. mossambica,* the *Tilapia zillii* was introduced from South Africa in 1963. On account of its small size, slow growth and aggressiveness toward other fish, it has never become popular with fish farmers in Taiwan.

In 1966, the Nile tilapia, *Tilapia nilotica,* was introduced from Japan. It and its hybrid with the *T. mossambica* have in a few years become very popular in Taiwan because of their rapid growth, large size and appealing appearance.

T. aurea was introduced in 1974 from Israel. The main purpose of the introduction was to obtain all-male offsprings by crossing it with *T. mossambica.*

Tilapia has become the most important freshwater pond fish in Taiwan in terms of quantity of production. In 1973, the total area of ponds in which tilapia is the main crop reached 4,528 ha, and the annual production of tilapia reached 10,923 mt.[2]

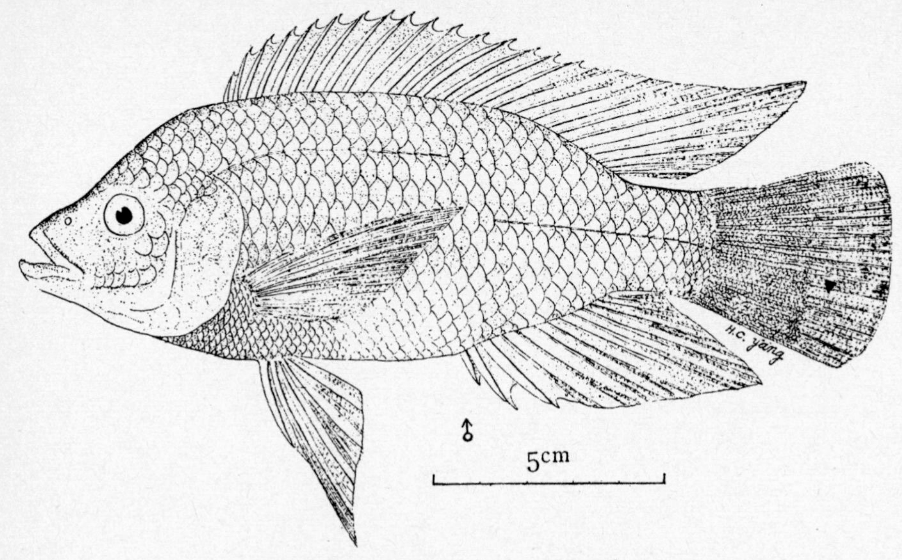

Fig 33 Java tilapia *Tilapia mossambica*

2. HABIT OF TILAPIA

2.1 Java tilapia, *T. mossambica*

This is the most commonly known of the tilapias. It grows to a fairly large size in Africa, reaching a length of 360 mm. In Taiwan, due to over-crowding in the ponds and the short growing period, it rarely attains 200 mm in length.

T. mossambica is a secondary freshwater fish and can thrive in brackish and salt waters. It is reared in the brackish water fish ponds of Tainan, where the salinity reaches 75‰ at times. Breeding also occurs in these ponds when salinity is below 30‰. A tropical fish, the minimum temperature it tolerates is reported to be about 10°C. Prolonged exposure to cold, however, even at temperatures above 14°C, will cause heavy mortality of the fish in ponds.

In Taiwan, the fish start to spawn when they reach the age of four months. The number of eggs from each spawning increases with the size and age of the fish, varying from 100 to over 1,000.

The sexes can be differentiated by the genitalia (3 orifices in the female and 2 orifices in the male). The males also have darker body colour and deep red fins and are larger than the females.

Before spawning, the pair of parent fish make a saucer-shaped depression (nest) on the pond bottom by digging with their mouths. The preparation of the nests, however, is not absolutely necessary for spawning to take place. Java tilapia have been observed to spawn in a pond with a concrete bottom.

In spawning, the female fish drops its eggs into the depression and is followed by the male, which ejects the sperm to effect fertilization. The fertilized eggs are immediately picked up by the female in its capacious mouth. The eggs are non-buoyant and non-adhesive. They are yellowish, elliptical and about 2·5 mm in the long diameter. In the mouth of the female, they are rolled around and are hatched in about 60 hrs at 28°C.

The larvae just hatched are about 4 mm in total length. They are feeble and remain in the protection of the mouth. In 3 to 5 days, the yolk sac is completely absorbed and the body length attains 8 mm. In another 2 or 3 days, they swim out from the mouth of the female fish.

The number of spawnings by one fish in one year is observed to be 6 to 11 in southern Taiwan. The interval between two spawnings is generally 22 days. The optimum temperature for spawning ranges from 20 to 35°C.

T. mossambica is not strictly herbivorous although its feeds are mainly of a vegetable nature. It feeds on plankton, algae, rice or wheat bran, soybean or peanut meal, chopped trash fish, etc. Live animal foods are accepted by the young fish, but not by the adults.

2.2 *T. zillii*

This species has never been popular with the fish farmers of Taiwan since its introduction. So virtually no study has been made of its habits. According to Breder and Rosen,[3] the *T. zillii* is strictly herbivorous, feeding on higher plants, and is aggressive towards other species. The optimum water temperature is 22–24°C, and 26°C is the best temperature for reproduction.

Breder and Rosen[3] stated that *T. zillii* spawn in typical cichlid fashion, on a clean stone or other smooth objects, and do not mouthbrood, although the parents do guard the eggs and young. They produce more eggs, up to 5,000 from a single large fish. This statement has not been confirmed by the experience of the fish farmers in Taiwan, who believe that the breeding habit of *T. zillii* is similar to that of the *T. mossambica*.

2.3 Nile tilapia, *T. nilotica*

The breeding habit of the *T. nilotica* is the same as the *T. mossambica*. It reaches 140 g in about 120 days and starts to breed at the age of five months. It is omnivorous, but feeds mainly on phytoplankton and higher plants.

2.4 *T. aurea*

As this species was introduced only recently, it has not been distributed to the fish farmers and no study of its habitat has been made.

3. POND CULTURE

3.1 Paddy culture

After its introduction in 1946, the *T. mossambica* spread widely in the southern part of Taiwan, but its extensive culture faced several difficulties:

a. Unfavourable climate – The fish is killed by the low winter temperature in natural waters in most parts of Taiwan. It has a short growing season and must be protected in the cold winter months.

b. Over-population – Due to its high fecundity, the Java tilapia tends to overpopulate the pond, and fish of small size are produced.

c. Low acceptance – The small size as well as the small edible portion of the fish affects its acceptance, except among the poor rural people.

In order to increase the supply of animal protein, especially for the poor people living in less accessible rural areas, the Government of Taiwan carried out a programme in 1951 to promote the paddy culture of tilapia and acquaint the farmers with the proper method of paddy culture developed by the Taiwan Fisheries Research Institute.[1] The programme was fairly successful at first. According to the Taiwan Fisheries Bureau, the acreage stocked with tilapia in 1952 was 3,438 ha of paddy fields and 1,906 ha of fish ponds. The average yield of fish in paddy fields was 235 kg/ha.

In spite of the initial success, the interest of the farmers in paddy culture of tilapia soon waned due to (1) the wide-spread use of pesticides on rice fields, which kill the fish, (2) the stealing of fish by people who consider the fish in the paddy as public property, and (3) shortage of labour. However, many fish farmers have learned the merit of the tilapia, ie fast growth, high production and small amount of care required, and the fish have become one of the favourite species for stocking polyculture ponds in southern Taiwan.

3.2 Utilization of sewage for tilapia culture[4]

One of the interesting developments is the utilization of sewage for tilapia farming. There are now some 30 hectares of fish ponds situated alongside the drainage canal in Tainan depending on the sewage-rich water of the canal to supply food for the tilapia. The sewage consists mostly of city waste and is of a black colour.

During the winter, the tilapia are moved into wintering ponds (same as wintering ponds for milkfish). At this time, the rearing ponds are thoroughly dried under the sun. Then the sewage water is let into the ponds and allowed to evaporate. After the pond bottoms become dry or partially dried, sewage water is again let in. This process is repeated 3 to 4 times before the ponds are again filled with water and stocked with tilapia from the wintering ponds in March. Generally no supplementary feeds are given, but sewage water is let in once every three days to make up the loss of water due to evaporation and seepage. The fish feed on the planktons and benthic algae, which grow profusely as a result of fertilization by the sewage water, as well as the detritus introduced with the sewage. Selective harvesting begins about 40 days after stocking and is continued at intervals of 10 to 15 days. The annual yield per hectare is 6,500 to 7,800 kg.

3.3 Stocking

The tilapia fingerlings may either be produced by the fish farmers themselves or procured from fish fry dealers. For monoculture or fish farming of which the tilapia is the dominant species (as in the case of fish-cum-ducks or fish-cum-hogs farming), the general practice is to plant 20,000 tilapia fingerlings of 2 to 3 cm in length in each hectare of pond. In polyculture with Chinese carps, the fish farmers generally stock each hectare of pond with 350 to 600 kg of tilapia fingerling of the size of 25 to 50 per kg. Most fish farmers now stock their ponds with the hybrid of *T. mossambica* and *T. nilotica*.

3.4 Pond management

When tilapia are polycultured with Chinese carps, no special management is required except multiple cropping or selective harvesting by netting to remove the fish that have reached marketable size of 200 to 300 g in weight. At this size the tilapia begin to propagate and the number should be reduced to avoid overcrowding. It is reported that the tilapia reproduce rapidly when the depth of water is about 40 cm, but, if the water exceeds 100 cm in depth and there is no shallow areas in the pond, they do not usually make nests and reproduction seldom occurs. Mono-sex culture of tilapia, although desirable for controlling their population, has not been practised in Taiwan.

Beginning in about 1972, culture of fish in combination with hog or duck farming has gained popularity in Taiwan. Many farmers have found rice farming to be unprofitable due to the low price of rice and high labour cost. They convert their rice paddies into fish ponds and build pigsties or duck houses beside the ponds. The excretions of the hogs or ducks are introduced into the ponds, with or without fermentation, to serve as fertilizers and/or feeds. Since that time, more than 5,000 hectares of paddy fields have been converted into fish ponds. Some reclaimed tidal land has also been used for fish-cum-hog culture. The fish stocked in these ponds are generally tilapia, silver carp, big head, grass carp and common carp, with tilapia (mostly *T. nilotica*) as the dominant species. It is estimated that 50 to 70 hogs or 2,000 ducks could supply sufficient fertilizers and feeds for the fish in a one-hectare pond.

3.5 Harvesting and marketing

Fish farmers harvest the tilapia of marketable size for sale many times in a rearing season to avoid overcrowding and to obtain money to finance their operation. So tilapia are harvested and marketed at nearly all times of the year. The largest number are put on the market, however, just before the onset of winter, when the ponds are drained and all the fish are removed. In some fish stalls in rural areas of southern Taiwan, one can find only tilapia and mullet in the months of December and January.

Tilapia are generally sold fresh, but live fish bring a better price.

3.6 The tilapia hybrid

In 1969,[5] the Lukang Fish Culture Station produced the hybrid of male *T. nilotica* × female *T. mossambica*, which was found to have an average daily growth rate of 1·16 g as compared with the 0·85 g of the hybrid of male *T. mossambica* × female *T. nilotica,* 0·74 g of pure *T. nilotica* and 0·59 g of pure *T. mossambica*. They call this hybrid Fu-shou Yu (Blessed fish) and use it extensively. In 1973, over 16 million fingerlings of this hybrid and their offspring (the F. hybrids are fertile) were produced by the Station and distributed to fish farmers. This resulted in an immediate increase in tilapia production.

The hybrid Fu-shou Yu is now immensely popular with fish farmers, not only because of its fast growth, but also because of its larger size, better colour appeal and consequently the much higher price it brings. However, the characteristics of faster growth and larger size are undoubtedly due to heterosis, and it is not possible to tell what the later generations will be like. Further experimentation is urgently needed.

LITERATURE CITED

1. Chen, Tung-Pai: The Culture of Tilapia in Rice Paddies in Taiwan. *Joint Commission on Rural Reconstruction Fisheries Series* No. 2. 1953.
2. Taiwan Fisheries Bureau: *Taiwan Fisheries Yearbook,* 1973.
3. Breder, C. M. & D. E. Rosen: *Modes of Reproduction in Fishes,* p. 495, 1966.
4. Huang, Ting-lang: Utilization of Sewage in Monoculture of Tilapia. *China Fisheries Monthly,* No. 182. 1968. (In Chinese.)
5. Kuo, Ho: Notes on Hybridization of Tilapia. Reports of Fish Culture Research Supported by Rockefeller Foundation. *Joint Commission on Rural Reconstruction Fisheries Series* No. 8. 1969.

VI. Culture of the Snakehead

1. INTRODUCTION

The snakehead commonly cultured in Taiwan (Fig 34), *Channa maculata* (*Ophiocephalus tadianus* Jordan and Evermann), occurs in creeks and ponds throughout Taiwan, especially in still water of about 1 m in depth with abundant aquatic plants. It possesses an accessory breathing organ similar to, though not as highly developed as, that of the labyrinth fishes, and thus they are able to withstand very low dissolved oxygen concentration as well as partial drying.

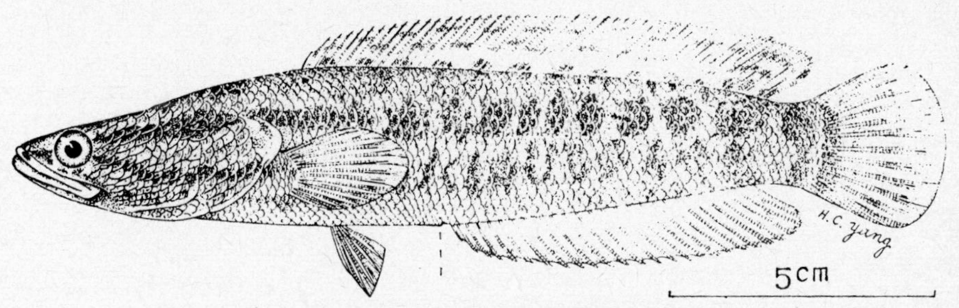

Fig 34 Snakehead *Channa maculata*

It is strictly a carnivorous fish, feeding on earthworms, tadpoles, shrimp, fish and other aquatic animals.

The snakehead in Taiwan matures in about two years.[1] The spawning season is from the middle of April to the middle of September, when water temperature ranges from 20 to 30°C. In natural water, the male and female fish make a slight depression on the bottom among the aquatic plants. Eggs are released at the break of dawn by the female while lying belly-up just above the depression. They are immediately fertilized by the male, which ejects the sperms while lying in a similar position. The fertilized eggs, about 2 mm in diameter, float among the aquatic plants until they hatch.

Fig 35 Natural water where snakeheads spawn

Fig 36 Newly hatched snakehead fry

The hatchlings measure 3·8 to 4·3 mm and are of a brownish colour. When they reach 9 to 10 mm in length, they start feeding on zooplankton and are able to fend for themselves. Up to this stage, the parents stand guard beneath the nest against enemies such as snakes, frogs and fish.

2. SNAKEHEAD FARMING

Until recently, the snakehead, on account of its predatory habit, was considered a pest fish, and its presence in a fish pond was regarded as highly undesirable. Not only was it not cultured, but it was removed whenever found. But since it has been found to have a high table quality and brings a good market price, fish farmers have become interested in its culture in recent years. Most fish farmers plant limited numbers in polyculture ponds, not only to reap some extra profits from the sales of this fish, but also to control the number of small wild fish as well as the fish (eg the tilapia) that tend to breed indiscriminately. A small number of farmers, however, put them under monoculture.

2.1 Supply of seed fish

Supply of seed fish is either by collection from natural waters (Fig 35) or by artificial propagation. To obtain them from natural waters, the fish farmers locate a nest of fertilized eggs or hatchlings, gather them with a small dip net and place them in a hatching basin or nursery tank, respectively. In the case of fertilized eggs, they could be placed in a large plastic basin with clean water. Aeration is not necessary, and any change of water should be gradual so as not to disturb the eggs. At a temperature of about 26°C, the eggs will hatch in about 36 hours, and at about 30°C, in 32 hours (Fig 36).

About three days after hatching, the yolk sacs are absorbed and the hatchlings can be placed in the nursery pond. The management practice in the nursery pond is the same as when dealing with hatchlings collected from natural waters. The nursery pond should be previously treated so that the water will contain plenty of natural food for the larval fish. The pond is first drained and the bottom exposed to the sun. Lime (1 kg per 3·3 m^2) is applied to neutralize the soil on the bottom and speed up the decomposition of the organic matters. Then fertilizers in the form of soybean meal, chicken droppings, etc. are applied and water let in. In about a week, there will be an abundance of daphnia and other plankton organisms in the water to serve as food for the larval fish.

Fig 37 Snakehead fry 12 days old

Sometimes daphnia and rotifers are cultured in a separate pond and scooped up to be used. The daphnia should be strained and only the small ones are fed to the larval fish. After about two weeks when the larvae assume an orange colour (Fig 37), chopped tubifex worms are fed, and in about 20 days unchopped tubifex worms are used. This ration, sometimes with the addition of chopped trash fish, is continued for six to seven weeks. By this time, they will reach 4 to 6 cm in length, turn dark grey in colour and may be stocked in the rearing ponds.

For induced spawning, mature fish of about two years old are used.[2] The female fish can be distinguished from the male by its lighter colour, soft and extended belly and the large pinkish genital pore (Fig. 38).

Fig 38 Female snakehead spawner

About 10,000 eggs can be obtained from a female of 1 kg, and over 30,000 eggs from a fish of 3 kg. The selected fish are kept in a brood pond two or three months in advance of the breeding season. They are given live food such as small fish and tadpoles. When warm weather sets in, say in March, they are ready to receive hormone injection, which varies in amount according to the maturity of the fish. Generally, for a spawner of 1 kg in weight, the pituitaries of one or several common carp of a total weight of 2 to 3 kg may be injected together with 20 rabbit units of Synahorin. The dose is divided into two equal portions, which are given at an interval of about 12 hours. The male fish receives in one injection only one-half of the dosage given to the female.[3] No second injection is necessary.

After the hypophyseal treatment, the practice in Taiwan is to place the injected fish into an ordinary fish pond for spawning and fertilization to take place. One male and one female fish are placed in a compartment of 3–4 m^2 formed by nylon nettings. Sometimes 5 or 6 pairs of fish are placed together in a small pond of 7 to 10 m^2 without segregation into compartments. Depth of water is kept between 60 and 100 cm. The top of the compartment or pond is covered usually with nylon netting to prevent the fish from jumping out. Plastic tanks of 0·5 to 1·0 ton capacity may also be used for this purpose. Spawning and fertilization usually take place the next day or so. The eggs that turn white are dead eggs and should be removed. The healthy eggs are of bright yellow colour, spherical, transparent, buoyant, non-adhesive, and about 2 mm in diameter. Development of the eggs and hatchlings is the same as when fertilized eggs are collected from natural waters.

2.2 Pond management

The snakehead are usually planted in polyculture ponds with Chinese carps.[4] Some fish farmers, however, put them in the same pond with tilapia, which serve as the forage fish.

2.2.1 Polyculture with carps

In case of polyculture with Chinese carps, they are planted for the purpose of eradicating the extraneous or pest fish. Not more than 500 snakehead of over 10 cm in length should be planted in a one-hectare pond. They may be planted any time between March and September, but only when the Chinese carps in the pond have exceeded 10 cm in length in order to avoid predation. It has been observed that a snakehead could easily devour a fish half its length. No special feeding stuffs are given; the snakehead forage on the wild fish and the young of the tilapia in the pond.

2.2.2 In combination with tilapia

In case of combination with tilapia, the ponds either have mud or concrete walls, but bamboo fencing or nylon screen of about 1·5 m in height should be erected on the top of the walls to prevent the fish from jumping out. The water should be 1 to 1·5 m in depth. Water hyacinth may be planted on the water surface to provide shade and a sense of security.

The initial stocking is 90,000 fingerlings of 10 cm in length per hectare. They must be sorted as to size two or three times a year and re-planted at a lower density to avoid cannibalism. The final stocking density is 15,000 to 24,000 fish per hectare. When only one or two ponds are available, the fish are segregated according to size in different compartments formed by nylon nettings.

A good arrangement is to have the tilapia confined to a section of the pond by nylon nettings of proper mesh size, through which the baby fish can swim out and be eaten by the snakehead. The disadvantage of this arrangement is the tilapia hatchlings are very small in size, and it takes many of them to make up a meal for the snakehead.

2.2.3 Monoculture

Monoculture of snakehead is not common, because the cost of production is high. The fish have to be trained to take artificial feed, which is usually a mixture of 80% minced trash fish and 20% formulated eel feed or wheat flour, which is less desirable. To train the fish to eat this artificial feed, the feed is first mixed with live daphnia. Then gradually the daphnia are eliminated.

3. HARVESTING AND MARKETING

If adequate natural feeds are available, growth of the fish is rapid in warm weather. The 10 cm fingerlings planted will reach 600 to 1,000 g in 9 or 10 months, or over one kg in one year. Survival may reach 90% if proper care is taken.

The snakehead are harvested at the close of the rearing season by draining the pond and capturing them by hand or with a dip net. They can be easily taken to the market alive by holding them in wet gunny sacks or any container with a small quantity of water. Only live fish are sold in the market.

LITERATURE CITED

1. Huang, Ting-lang: Aquaculture in Taiwan. *Quarterly Journal of Bank of Taiwan,* Vol. 25, No. 1, 1974. (In Chinese.)
2. Huang, Tsing-man: Artificial Propagation of the Snakehead. *China Fisheries Monthly,* No. 236, 1972. (In Chinese.)
3. Lee, Sheng-liang: A Preliminary Report on Artificial Propagation of *Channa maculata* (LACÉPÈDE). *Aquaculture,* Vol. 1, No. 3, 1971. (In Chinese.)
4. Lee, Hsiao-chao and Sheng Hsiang Chen: The Habit and Culture of Snakehead. *China Fisheries Monthly,* No. 232, 1972. (In Chinese.)

VII. Culture of Walking Catfish

1. GENERAL STATUS

The walking catfishes, *Clarias fuscus* and *C. betrachus*, are considered a pest in the United States and some western countries due to the fact that they are predatory fish and that they are capable of 'walking' from one pond to another, but it is a commercial food fish in South Asian countries. In Hong Kong and Taiwan, the Chinese believe that the walking catfish, steamed or prepared in soup, has tonic quality and it commands a high price in restaurants.

The indigenous walking catfish in Taiwan is *C. fuscus,* also known as white-spotted freshwater catfish (Fig 39). Around 1972, the *C. betrachus* was introduced from Thailand. The former is smaller in size, generally weighing 150–200 g and reaching a maximum size of about 500 g. The latter is faster growing, reaching a weight of two kilograms in a year. The *C. fuscus*, however, is still the favourite among consumers and fish farmers. Its culture is described herewith.

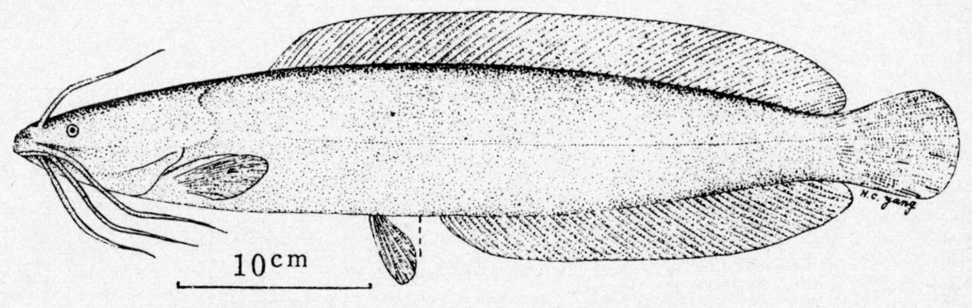

Fig 39 Walking catfish *Clarias fuscus*

The *C. fuscus* is found in all streams, irrigation ditches and ponds. It is a slimy fish with strong sharp spines and, therefore, difficult to handle. Because of its relatively small size and the difficulty of dressing the fish without suffering painful wounds, its popularity is more or less limited to restaurants and rural households.

Until about 1969, it was not cultured in ponds, but caught in the wild. With increasing industrial pollution and widespread use of pesticides by farmers, its natural supply has diminished. As a result of its high price, culture of the fish has gained impetus and is now practised by many fish farmers in different parts of the Island.

2. ARTIFICIAL PROPAGATION

Because of the increasing difficulty of obtaining the fingerlings from the wild, the Taiwan Fisheries Research Institute attempted the artificial propagation of *C. fuscus* in 1970 and succeeded in obtaining a survival rate of 80–90% on the 40th day after hatching.[1]

To get the best result, the spawners should be cultivated in ponds of 35–350 m^2.[2] The walls of the pond should be of concrete or brick and the bottom of mud. The depth of water should be 1–1·5 m. To provide hiding places for the fish and to facilitate catching the fish, caves should be provided along the sides, or pipes made of cement or plastic may be placed on the bottom. Floating aquatic plants such as water hyacinth and duckweed should be placed over one half of the water surface to provide shade and a sense of security to the fish.

Fish of over 1·5 years of age (over 150 g in weight) should be placed in the spawners' pond 3–4 months in advance. It is best to have the sexes separated by ponds. The best stocking rate is 15–17 fish per m^2.

The feeds consist of minced trash fish, animal viscerals, rice or wheat bran and soybean or peanut meal in the amount of 3–5% of the total weight of fish. The amount should be reduced in case of a change in water quality or a lower water temperature.

If the spawners are from a wild stock, they should receive immediate hormone treatment to avoid the ill effect of injury and reabsorption of the sperm and eggs during long holding.

Maturity is generally reached in fish at $1\frac{1}{2}$ – 2 years of age. The female will have soft extended belly and round, enlarged red genital (Fig 40). The belly of the male fish is flat and the genital pore small and oval in shape.

As in the case of Chinese carps, the pituitary of common carp and Synahorin, Gonagen or Pubergen are injected to induce spawning. The dosage is pituitary from a carp of 2–3 times the body weight of the recipient catfish. For female catfish of 190 g or less in weight, 20 rabbit units of Synahorin or 120 international units of Pubergen are injected with the pituitary. The mixture is suspended in Ringer's solution (0·7% NaCl,

0·03% KCl, 0·026% CaCl and 0·003% $NaHCO_3$) and divided into two portions to be injected into the female fish at an interval of 8–10 hours. Injection is intramuscular below the pectoral fin.

Large and older males need no injection, but the small and younger male fish need to be injected with about one half the total dosage given to the female at the time when the second injection is given to the latter.

After the hormone treatment, the male and female fish are impounded in large plastic tanks or small concrete ponds with adequate aeration. At a temperature of 26–29°C, they will be ready for stripping in 18–24 hours, depending on the difference in temperature and the degree of the maturity of the fish.

To determine the ripeness of the eggs, the belly of the female fish is gently pressed. The ripe eggs will flow freely and are dark red in colour. The green eggs or the over-ripe eggs cannot be stripped easily and are of whitish colour. Stripping the female fish follows the same procedure as in the case of the Chinese carps. But the male fish cannot be stripped and must be cut open to obtain the sperms. The testes are symmetrically located on the two sides of the abdominal cavity. They are pinkish, elongated and serrated along the margin. They are removed, cut into strips of about 3 mm in width and mixed thoroughly with the eggs to effect fertilization. The fertilized eggs, 0·19 mm in diameter, are washed 5 or 6 times with water and put into the hatching pond.

Fig 40 Female walking catfish and gonads

Fig 41 Walking catfish caught in plastic pipe and emptied into container

The hatching ponds are small concrete or brick ponds with water of about 60 cm in depth. The fertilized eggs are evenly spread on the surface of nylon screens in trays of wooden or wire frame and submerged just below the surface of the water. The fertilized eggs are adhesive after absorbing water and become attached to the screen.

At a water temperature of 27–29°C, the eggs hatch in about 30 hours. The hatchlings are about 0·46 mm in length, with large yolk sacs of about

0·18 mm in width. They lie on the screen and are inactive excepting the wiggling of the tails. After three days, the yolk sacs are absorbed. The fry become mobile and start feeding.

During hatching, adequate dissolved oxygen should be provided either by aeration or gradual change of water, without disturbing the eggs or hatchlings. Sudden change of water temperature should be avoided. To minimize mould infestation, 0·2 ppm Malachite Green or 1–2 ppm Methylene Blue may be applied to the water.

3. REARING OF THE FRY

When the fry start feeding on the fourth day, they are fed daphnia and rotifers. These zooplanktons must be supplied in large quantities, and those larger than 600 microns should be filtered out.

From the eighth day on, they may be fed tubifex and chopped oyster or clam meat in addition to daphnia. beginning from the 30th day, they are fed minced fish, animal blood and viscerals mixed with rice bran, wheat bran, soybean meal or peanut meal, etc. On the 40th day, they would have reached 2·5–4·0 cm in length, a size suitable for stocking the ponds. The survival rate may be as high as 70%, barring mishaps.

If the fry are moved to earth ponds when they are capable of ingesting tubifex, better growth rate may be obtained. But they must be protected from predators and the bottom of the pond should be drained and dried under the sun before water is let in.

Since daphnia form the most important feed of the catfish fry, they should be cultured in a separate pond. The earth pond is first drained and the bottom dried under the sun. After application of lime (300 kg per 0·1 ha) and fertilizers (barnyard manure or inorganic fertilizers), water is introduced to a depth of 60 cm. Daphnia will develop in abundance after about two weeks. They may then be collected with a dip net and fed to the hatchlings.

4. REARING FOR MARKET

4.1 Pond construction and stocking

Since the walking catfish has the habit of boring its way through mud banks, the walls should be of concrete or brick. If the banks are of mud, they should be thick and strong. The ordinary size of the pond varies from 80 to 1,600 m^2, with banks about 1·5 m in height. The water depth is maintained at 40–100 cm. A number of caves are provided along the banks or pipes (concrete or plastic) of about 13 cm in diameter and 5 m long may be sunk to the bottom to provide hiding places for the fish. As in the case of the spawners' pond, about one half of the water surface should be covered with floating aquatic plants to provide shade.

Monoculture of walking catfish is a recent practice in Taiwan. Since the fish has high tolerance for oxygen deficiency, the stocking rate is high. Generally 100–200 fingerlings of 2·5–3·5 cm in length (total weight 1–1·5 kg) may be stocked in an area of 3·3 m^2. As the fish increase in size, they are redistributed into ponds at lower stocking rates.

4.2 Feeding

The walking catfish is an omnivorous fish and may be fed either animal or vegetable food. One efficacious feed consists of a mixture of minced fish, soybean meal powder and cooked oatmeal formed into balls. Cracked snail may also be given. The feeds are put into a wire basket and lowered into the water. Usually the fish are fed once a day in the late afternoon. The amount of feeds is from 3–5% of the total weight of the fish, varying with weather, water condition and feeding intensity.

4.3 Harvesting and marketing

Growth of the fish is most rapid in the period from the third month after hatching to one year old, after which growth slows down. Fingerlings of 2·5–3·5 cm reach a weight of about 120 g in one year.

Fig 42 Walking catfish can be transported long distances in plastic bags partially filled with oxygen

The fish is marketable when it exceeds 120 g in weight, but the larger the size the better the price it brings. A pond of 0·1 ha will produce about 2,160 kg of fish of 150 g in individual weight per year. The walking catfish is sold live. To meet the market demand from time to time, selective harvesting is practised. The fish may be caught by dipnets from the caves where they hide or by lifting up the pipes and emptying them into a net (Fig 41). Those of marketable size are separated from the undersized fish, which are returned to the pond for further growth.

After capture, the fish are impounded in small cement or brick ponds for one or two days to empty the guts and rid them of the muddy taste. They are then taken to the local market in metal or plastic tubs. For longer distances, they are transported in plastic bags partially filled with oxygen (Fig 42). In the market, they are kept alive in tanks with aeration.

LITERATURE CITED

1. Huang, Ting-lang *et al*: Experiment on Artificial Propagation of the White-spotted Freshwater Catfish, *Clarias fuscus* (Lacepede). *JCRR Fisheries Series* No. 12, 1972. (In Chinese with English abstract.)
2. Huang, Ting-lang: Aquaculture in Taiwan. *Quarterly Journal of Bank of Taiwan*, pp. 156–165, 1974. (In Chinese.)

VIII. Culture of the Mud Skipper

1. INTRODUCTION

The mud skipper, *Boleophthalmus chinensis* (Fig 43), is an edible brackish-water goby. It is a delicacy among the people of Taiwan and is the highest priced fish locally.

It is a small fish. The adults generally measure 12 to 15 cm in total length, weighing 30 to 40g.[1] No record of its production in Taiwan is available. It is certain, however, that only a small quantity is marketed. Although quite common in the sea food restaurants of southern Taiwan, it is seldom served in the restaurants of Taipei.

The mud skipper is found on mud and sandy flats in the tidal zone, particularly on the western and southern coasts of Taiwan. It builds a tunnel with two or more entrances, which serves as shelter as well as the spawning chamber. At low tide, it leaves the tunnel and slides on the sandy or rocky surfaces to feed on the benthic algae, mostly diatoms.

All supply of mud skippers used to come from catching them in their natural habitat. Because of the difficulty of capturing the fish and the high price paid by the restaurants, as much as US$8·00 per kilogram,[1] many fish farmers have become interested in their culture. At the time of writing, some 40 hectares are devoted to its culture in the Tainan and Yunlin areas.

2. POND CULTURE

2.1 Supply of seed fish

The season of reproduction of the mud skipper is from April to September. Most of the seed fish are captured in the months from June to August. They can be found in puddles on the mud or sandy shores along the estuaries at ebb tide. They are generally 1·5 to 3·0 cm in length and occur most abundantly in areas south of Yunlin in brackish water of less

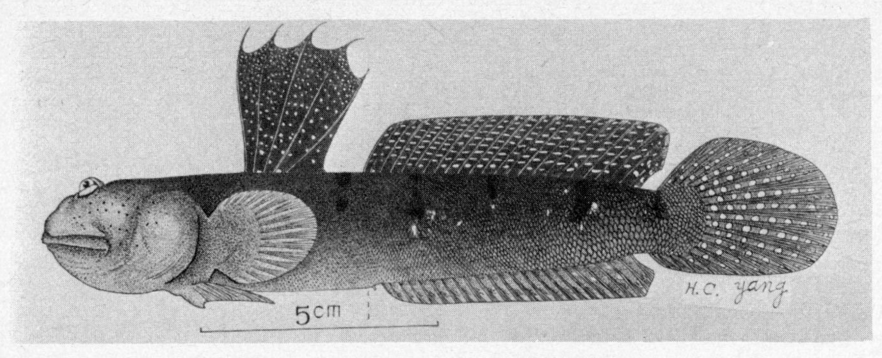

Fig 43 Mud skipper *Boleophthalmus chinensis*

Fig 44 Mud skipper pond

than 1·015 specific gravity. The fry collectors catch them from the puddles with small dip nets and sell them to the fish farmers.

2.2 The fish pond

The ponds for mud skipper culture are generally small, varying from 0·1 to 1·0 hectare in size (Fig 44). They are substantially the same as milkfish ponds, except that bamboo or plastic fences should be erected on the dykes to prevent the escape of the fish and the entrance of predators. Since it is necessary to drain the pond and sun the bottom, the elevation of the pond bottom must be above the mean water level. Ditches of about two metres in width sloping toward the sluice gate are provided in the middle of the pond and along the sides to facilitate drainage and serve as shelter for the young fish during hot sunny days.

The pond bottom should be of loam, because benthic algae grow better on loamy soil and the tunnels excavated in loamy soil do not collapse as easily as those built in sand.

The specific gravity of the water is preferably kept at 1·010 to 1·020.

2.3 Pre-stocking treatment of the pond[2]

Since the mud skipper feeds on benthic algae, it is necessary to obtain a good growth of the latter before the ponds are stocked. This is accomplished by sunning the bottom and the application of fertilizers, as is done to the milkfish ponds. There are however, some differences in the practice: (1) The algal pastures need not attain such thickness, so less fertilizer needs to be applied. (2) No repeated introduction of sea water and evaporations are done, as they will increase the salinity of the water. For each hectare of newly excavated ponds, about 600 kg of barnyard manure and rice bran are applied, and sea water of low salinity is introduced to a depth of 15 cm. The algal pasture will consist of diatoms and blue-green algae.

Snails and Chironomid larvae are consumers of the benthic algae and should be eradicated by the application of 0·3 ppm Bayluscide for the former and 0·25 ppm Abate, 0·3 ppm Sumithion or 0·25 ppm Lebaycid for the latter.

2.4 Stocking the pond

Mud skipper fingerlings are obtained from the professional collectors. The stocking rate is generally 30,000 fingerlings per hectare, with a

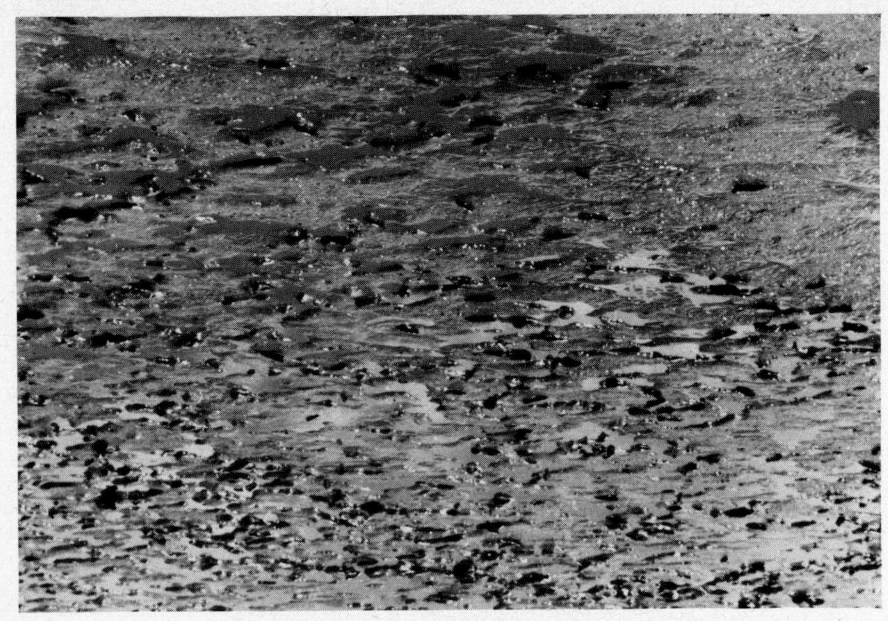

Fig 45 Drained mud skipper pond showing tunnels

Fig 46 Mud skippers marketable size

maximum of 50,000. Since it is difficult to recapture them once they are planted in the pond, segregation as to size is not done. The mud skipper is not cannibalistic anyway.

2.5 Pond management

Management of a mud skipper pond is a modification of the practice in milkfish culture. Since both species feed on benthic algae, the water must be kept clear and shallow to allow good penetration of sunlight to the bottom. At the initial stage, the water in the pond is maintained at about 15 cm in depth (about 30 cm in the ditches). The mud skippers are then still small in size and make only shallow tunnels, so the pond is not treated. After about 45 days, when the fish exceed 5 cm in length, they dig deeper tunnels in which they could hide, and the algal beds have become too decimated to support the growth of the fish. At this point, the pond is drained (except that some water should be left in the ditches) and dried under the sun for 3 to 6 days (Fig 45), after which fertilizers in the form of nightsoil, rice bran, etc. are applied. Sea water of 1·010 to 1·020 specific gravity is then introduced, and a new layer of benthic algae will form.

During this treatment, the mud skippers will hide in their tunnels. Care must be taken so that no large amounts of nightsoil get into the tunnels and kill the fish. Mixing the fertilizers in the incoming sea water is safer and will accomplish the same purpose. After this, the water depth may be reduced to 2 to 7 cm.

The growth of the mud skipper is best at temperature above 28°C. It ceases to grow at a temperature below 14°C.

2.6 Predators

The common predators are fish (tilapia, *Elops* sp., etc.), which should be removed when the pond is drained and screened out during the introduction of water, birds (which should be frightened off) and crabs (which should be kept out with fences).

3. HARVESTING AND MARKETING

The mud skipper takes one to two years to grow to marketable size (Fig 46), depending on the condition of the pond and the management. The smallest marketable size is 24 g, and the largest may reach 40 g. Survival rate is about 60%.

Harvesting is done by covering the main entrance of the tunnel with a bamboo trap (Fig 47). The job is done by a professional mud skipper catcher. The mud skippers may also be caught with a dip net placed at the water inlet. When water is being introduced, the mud skippers gather just above the net and may be caught by lifting the net.

Transactions are done between the fish farmers and the dealers at the pond side. Mud skippers can be transported over long distance by simply holding them in a small amount of water at the proper temperature.

4. ARTIFICIAL PROPAGATION[3]

Experiments on the artificial propagation of the mud skipper were carried out by the Tungkang Marine Laboratory of the Taiwan Fisheries Research Institute in the summers of 1972 and 1973. The brood fish used were 10 to 20 cm in length. They were placed in plastic tanks and provided with plastic pipes for them to hide in. Enough water was put into the tank to cover just the nostrils of the fish.

The females were injected with $\frac{1}{2}$ to 1 carp pituitary and 5 to 10 rabbit units of Synahorin. A second injection was given on the third day. On the fourth day, they were ready for stripping. One female yielded over 10,000 eggs. The eggs were spherical and yellowish in colour, measuring 0·5–0·6 mm in diameter. They were demersal and adhesive.

Sperms were obtained by cutting open the males, taking out the testes and cutting them into small pieces, which were then thoroughly mixed with the eggs. The fertilized eggs were washed several times with sea water of 14·9–18·5‰ salinity.

The fertilized eggs hatched in 65 hr and 40 minutes at water temperature of about 28°C. The larvae at hatching were about 2·8 mm in length (Fig 48). They all died on the fifth day without any significant increase in length.

LITERATURE CITED

1. Huang, Ting-lang: Aquaculture in Taiwan. *Bank of Taiwan Quarterly Journal,* Vol. 25, No. 1. 1974. (In Chinese.)
2. Ting, Y. Y. and M. N. Lin: Farming of the Mud Skipper, *Boleophthalmus chinensis* (Osbeck) in Taiwan. *China Fisheries Monthly* No. 238. 1972. (In Chinese.)
3. Liao, I-chiu, Nai-hsien Chao, Lei-chiang Tseng and Shang-ching Kuo: Studies on the Artificial Propagation of *Boleophthalmus chinensis* (Osbeck) – 1. Observation on Embryonic Development and Early Larvae. *JCRR Fisheries Series* No. 15. 1973. (In Chinese.)

Fig 47 Bamboo trap for catching mud skippers

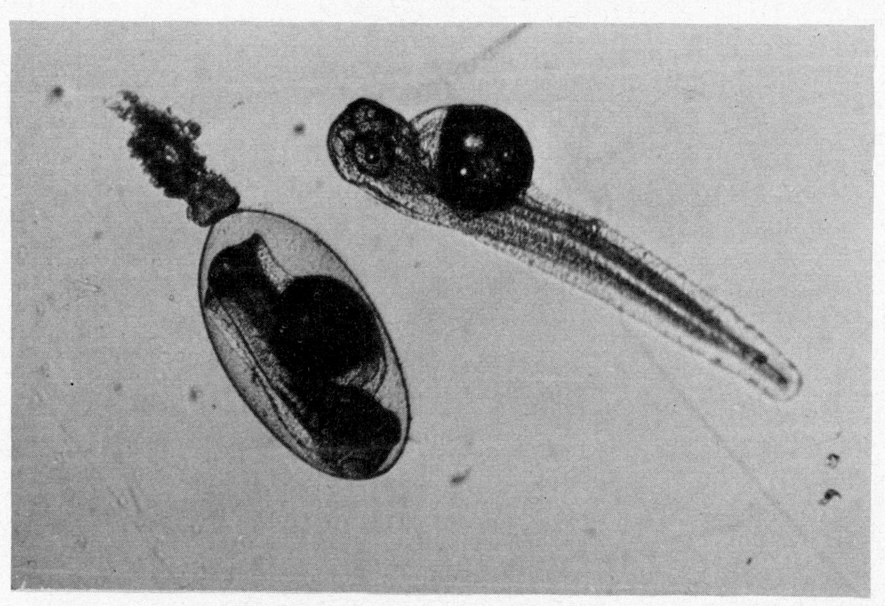

Fig 48 Egg and hatching of mud skipper

IX. Oyster Culture

1. INTRODUCTION

Fourteen species of oysters have been identified in Taiwan: *Crassostrea gigas, C. echineta, Lopha cristagalli, Dendostrea folium, Ostrea turbinata, O. crenulifera, Saxostrea dubia, S. vitrefactor, S. paulucciae, O. denselamellosa, Caxostrea mordax, Pycnodonta chemnitzi, P. hyotis,* and *P. hyotis imbricata.*[1] The Japanese oyster, *Crassostrea gigas* (Fig 49), however, is the main species cultured. The strain is found to be similar to that cultured in Kiushiu of southern Japan.

Culture of the oyster began in Taiwan probably some 200 years ago. According to the Fisheries Year Book[2] published by the Taiwan Fisheries Bureau, the area of oyster beds and production in 1973 are shown in Table 4.

Table 4. Area of oyster beds and production in 1973

	Area (ha)	Production (mt)*
Hsinchu Hsien	447	2,052
Taichung Hsien	339	712
Changhua Hsien	2,350	4,276
Yunlin Hsien	3,670	3,239
Chiayi Hsien	1,844	3,215
Tainan Hsien	325	596
Kaohsiung Hsien	480	71
Pingtung Hsien	10	7
Penghu Hsien	1	32
Tainan City	74	89
Kaohsiung City	6	39
Taoyuan Hsien	6	15
Total	9,552	14,348

*Weight of shucked oyster

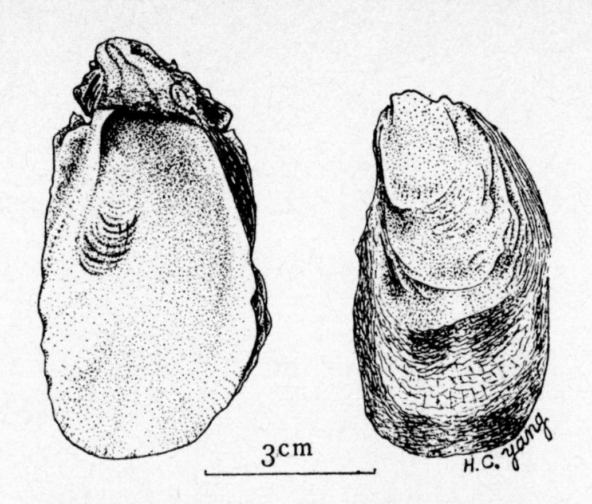

Fig 49 Japanese oyster *Crassostrea gigas*

Fig 50 Sticks bamboo (right) and plastic (left) laden with oysters

Oyster was once the food of the poor people in Taiwan. The price of oyster meat in 1963 was only NT$6 to 10 (US$0·15–0·40) per kg. In recent years, due to increasing demand from sea food restaurants as well as common households and the rise of labour cost, the retail price of oyster (shucked) has risen to NT$60 (US$1·50) per kg. The income of the oyster growers has consequently been much increased.

Oyster culture contributes to the coastal fisheries resources in another sense. The shallow waters on and around the oyster beds are rich in food organisms (including the eggs and larvae of the oyster). They therefore serve as nursery grounds for young fish and crustaceans.

As seen from Table 4, oyster farming in Taiwan is confined to the south and southwestern coasts where large stretches of sandy bottoms are available for the planting of bamboo sticks to serve as cultch.

2. ENVIRONMENTAL CONDITIONS

In some respects, the environmental conditions of the coast of Taiwan are favourable for oyster farming. First, the year-round high temperature of the sea water (15–30°C) speeds up the growth of the oyster and enables spats to be collected nearly all time of the year and in nearly every part of the coastal waters. Secondly, there are abundant nutrients in the coastal waters as a result of discharges from rivers and drainage canals on the western coast. Thirdly, the culture of oyster on bamboo sticks is a labour consuming practice, and Taiwan, up to the present, can provide the cheap labour required.

These seeming advantages, however, have their drawbacks. The year-round high temperature enables the oyster to spawn at all times of the year along the western coast. This makes it impossible to control the number of spats on the cultch to avoid overcrowding. There is also no separation of the oyster beds into spat collection, growing and fattening grounds. Another disadvantage of the year-round spawning is that the spawning activity dissipates a large part of the energy that could otherwise be spent for growing. The discharge of water from rivers and drainage canals creates the problem of water pollution, which may be the cause of mass mortality of oysters and clams in recent years. Cheap source of labour, mostly family labour at present, may not be readily available in the near future with the recent rapid development of industry and the consequent exodus of young able-bodied people from the rural areas.

On top of these disadvantages, the attacks of typhoons and floods often create havoc on the oyster beds by washing away the oyster-laden bamboo sticks and oyster racks.

3. CULTURAL PRACTICE

3.1 Collection of spats

Bamboo sticks serve as cultch on which the oyster grow to marketable size at most places in Taiwan. Only at a few places (Tainan and Chiayi) where there are sheltered areas with a water depth over 1 metre is off-bottom cultch used.

Since there is no separation of spat collection grounds from the growth and fattening grounds in most oyster growing areas, the oyster growers simply plant bamboo sticks on the shallow bottoms. On these sticks, the oyster spats are collected and the seeds are allowed to grow to marketable size.

The sticks used are of different bamboo species. For durability, they should be from older bamboo trees, and they should be low in cost. It is now the practice in some areas to treat the sticks with creosote to prolong their service life. It is said that treated bamboo sticks may last five years or more, but untreated sticks will last for only one or two years. Damage is mostly from the shipworm, *Teredo*. Plastic sticks have been tried, but their cost is prohibitive (Fig 50).

In most cases, the bamboo sticks are split from bamboo poles. Their length varies from 30 to 95 cm and width from 1 to 5 cm according to the depth of water at the locality. In the areas south of Yunlin, the bamboo sticks are split at the upper end to hold one or two oyster shells as collectors.

Because the peak of the oyster spawning season in Taiwan is in December, the bamboo sticks are generally planted during the period from October to February. Too late planting will catch less spats and also delay the growth of the oyster, while too early planting will collect more barnacles and cause more fouling of the sticks. Sticks are also planted in July and August to collect what are called 'fall spats'.

Collection of spats is not possible in the Pingtung Hsien due to unfavourable salinity and Kaohsiung City areas due mainly to heavy pollution. The oyster seeds required in these areas are supplied from the Kaohsiung County and Tainan areas, where bamboo sticks with oyster seeds on shells are trucked and sold to oyster growers in the former two areas for planting in March and April.

In 10 to 15 days after their deposition, the oyster seeds attain a size of 2 mm and are visible to the naked eye. They are sold and shipped when they reach a size of 6 to 9 mm. The number of seeds on a shell varies a great deal, but the ideal number is 25 to 30.

In recent years, spats have been collected along the coast of Changhua and shipped to Penghu, where spat collection is difficult. These spats are collected on oyster shells strung on a plastic rope stretched horizontally between two bamboo sticks. About 150 shells are strung on a rope of three metres long and are spaced by loops on the rope.

3.2 Culture on bamboo sticks

The bamboo stick method of oyster culture is the prevailing method in Taiwan. This is due to its simplicity and low investment as well as the fact that the western coast of Taiwan consists of gently sloping shallow sandy bottoms not suitable for the broadcasting off rocks and off-bottom culture. Any area where the seawater has a specific gravity of 1·005 to 1·025 (optimum 1·020), at an elevation of less than two metres above low tide, and with an exposure time not exceeding the submerged time can be utilized for this method of oyster culture.

The bamboo sticks are planted in rows (Fig 51). The space between rows is 40 to 80 cm and that between sticks is 20 to 30 cm, varying with localities. At every 4 to 12 m, a passage of 50 to 100 cm in width is provided to facilitate working on the beds. Passages of 2 to 3 m in width are also provided at suitable places to allow the entrance of bamboo rafts for transportation of the cultch material and the harvested oysters.

After the bamboo sticks (with oyster spats) are planted, the oyster growers have to make frequent inspections (sometimes daily) of the beds and make whatever adjustments that are needed. The disarrayed or fallen sticks are put back into place. Some sticks may need to be raised to avoid being buried by sand and allow the oysters on the lower end to grow. This is especially needed after a typhoon or flood. The oysters that have fallen from the sticks are picked up to be sold. Oyster drills (adults and eggs) and other pests are also removed and eradicated when they are recognized. The Changhua County offers a reward for catching the oyster drills and their eggs.

After 7 to 8 months, the oysters reach a length of 4 to 7 cm and can be marketed. Some harvesting of the oyster takes place at all times of the year in Taiwan, but heavy harvesting takes place in the period from July to September and the period from November to February, when the oysters are fatter.

Harvesting is done either by breaking off the oysters from the bamboo sticks with a steel pick and picking up the fallen oysters on the ground or by pulling up the entire sticks. These are loaded on to bamboo rafts and

Fig 51 Rows of bamboo sticks

Fig 52 Wind-driven cart for transporting oysters

taken to an open area in the village to be shucked. Sometimes a wind-driven cart is used to transport the oysters (Fig 52).

Shucking is done by women, usually members of the oyster grower's family. The yield of meat varies according to the size and condition of the oyster. The lean oyster will yield only 7 to 9% of meat, but the large fat oyster will yield as high as 20%. The average is 10–15%. Bloating with fresh water to increase the weight of the oyster is universally practised.

The oyster meat is sold to travelling buyers who come to the growers to collect the meat and sell it to fish dealers, peddlers or processors who produce dried oysters.

The oyster shells are sold to be used for making quick lime.

3.3 Off-bottom culture

Off-bottom culture of oyster is comparatively new in Taiwan and is practised mainly in Chiayi, where nearly 80% of the oyster production is from off-bottom culture, and in Penghu (Pescadores Islands), where all oyster production is from off-bottom culture (Fig 53).

Both raft culture and long line culture are practised in Taiwan. Raft culture is more common in deeper water further away from the shore. The rafts are made up of bamboo poles and are usually 15 m × 7·5 m. They are usually buoyed by empty oil drums and anchored to the bottom. Plastic strings, instead of the wire used formerly, with oyster shells strung 15 to 20 cm apart are suspended from the rafts. These shells are spaced by loops made on the strings. At most places, the shells serve as spat collectors as well as for growing the oysters for market. But, in some places, notably in Penghu, shells with spats collected in Changhua and Chiayi are used. Very little care is required during the growth of the oysters. The oyster farmers need only to inspect the rafts frequently to see if there is any damage to them and to remove those oysters of suitable market size, particularly if the strings become too heavy.

Long line culture is practised in shallower water, and is especially common in estuaries and drainage canals (Fig 54). A plastic line of about three metres long is stretched between two bamboo sticks, and about 150 oyster shells are strung on the line. As in the case of raft culture, a hole is made in each shell and the plastic line runs through it. The shells are spaced by loops made on the line. In most cases, the shells serve both for spat collection and growing the oysters. Different from the raft culture method, the long line is exposed to air part of the time.

According to Lin,[3] in Chiayi the oyster grows fast in the period between April and August. Growth slows down or stops from September

Fig 53 Off-bottom culture of oysters in estuary

Fig 54 Long-line culture of oysters in estuary

to January. In a period of nine months, the oyster grows to 6–7 cm, a growth condition similar to that in Hiroshima, Japan.

According to experiments of long line culture in Suao Bay (Ilan) and Makung Harbour (Penghu), oyster seeds of 8 mm in length grow to 6·8 cm in four months.[4]

In raft culture, the oyster farmers harvest their oysters by going out to the oyster beds in bamboo rafts (with or without engines), removing the lines laden with oysters, and taking them back to shore, where they are culled and shucked by women. The oyster meat is also bloated by soaking it in fresh water, which will increase the weight by 30%.

In Taiwan all oysters are eaten in cooked condition for the reason that most oyster growing areas are heavily polluted and that oysters have spawns at most times of the year.

4. MASS MORTALITIES

In recent years mass mortalities of oysters and clams have occurred along the western coast in the months of April and May. The loss has been variously estimated to be about 30% of the number planted. It has been reported in Chiayi County that the mortality did not affect the production of oysters. The removal of the dead oysters reduced the density of the population on the beds and resulted in better growth.

The cause of the mortality is as yet undetermined. Pollution from industrial waste, high density of planting, change of salinity or temperature, or disease have all been suspected. Investigation of the cause and study of control measures are underway.

LITERATURE CITED

1. Kuo, Ho: Economic Molluscs of Taiwan. *JCRR Special Publication* No. 38. 1964. (In Chinese.)
2. Taiwan Fisheries Bureau: *Taiwan Fisheries Year Book,* 1973.
3. Lin, Yao-sung: Biological Study of Oyster Culture in Chiayi. JCRR Publications. *Fisheries Series* No. 8, pp 77–115. December 1969. (In Chinese.)
4. Huang, T. L.: Aquaculture in Taiwan. *Journal of Bank of Taiwan,* Vol. 25, No. 1. March 1974. (In Chinese.)

X. Culture of Clam, *Meretrix lusoria*

1. GENERAL STATUS

The *Meretrix lusoria* (Fig 55) is found along the western coast of Taiwan and in scattered spots along the northern coast wherever the sandy beaches slope gently toward the sea. It inhabits bottoms with a sand content of 50–90%, especially those of 60–80% sand. The specific gravity of the sea water should be 1·010–1·024. According to the Taiwan Fisheries Year Book,[1] the total culture area amounted to 3,799 ha, with a production of 4,283 mt in 1972. Its culture is either for seed production or for rearing to market size.

2. SEED PRODUCTION

Clam culture in Taiwan began as early as 1925 in Kaohsiung Harbour and gradually spread to other areas along the western coast.[2] Its expansion was formerly limited by the supply of seed clams. Although clams are found practically along the entire western coast, seed clams were captured only in a few localities. Up until about 1970, the seeds were supplied mainly from the estuary of the Tansui River of Taipei Hsien and, in smaller quantities, from Peimen of Tainan Hsien. There the seeds were collected with rakes operated from small crafts. The size of the seeds varied from 700–1,000/kg. Although they were found the year round, the main season of collection was from May to August.

It was around 1971 that the fish farmers began to produce seed clams of stocking size. The practice has spread. In the Chiayi area some 700 ha of milkfish ponds are now used for this purpose. The fish farmers gather the tiny clams with a sieve from the sandy flats in tidal areas. These tiny clams are about 0·5 mm in length and white in colour (Fig 56). They are sold as a mixture of clams and sand. One kilogramme of this mixture usually

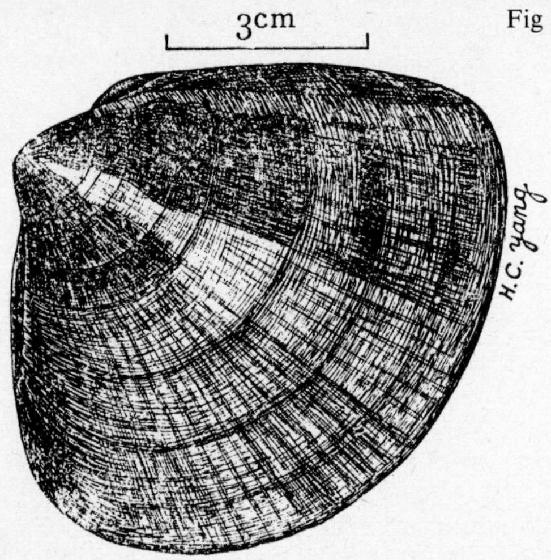

Fig 55 Taiwan hard clam *Meretrix lusoria*

Fig 56 Tiny seed clams

contains about 30,000 tiny clams. The higher the content of clams, the higher will be the price of the mixture.

The tiny seed clams are sold to seed clam growers, who put them in shallow (30–50 cm in depth) brackish-water ponds. If the pond is of sufficient fertility, no application of fertilizers is necessary.[3] Otherwise, fertilization is desirable. The fertilizers applied are nightsoil, hog manure, chicken droppings, rice bran, etc. The pond water should be clear and of light brown colour at 1·010–1·017 specific gravity. If this is not the case, the situation should be corrected either by change or addition of water. Generally about one half of the pond water is changed every three or four days. Fertilizers are applied during the change of water.[4]

Stocking rate varies considerably according to management practice and environment of the pond, but generally it ranges from 30 to 50 million per hectare.[3] They are either scattered as evenly as possible over the entire pond or concentrated to cover one half of the bottom. In the latter case, they are sifted as to size (Fig 57) and re-stocked in different sections of the entire pond at a later stage. Bamboo sticks are planted in the different sections to serve as markers indicating the time of stocking and size of the clams.

During the sifting, the miscellaneous species of snails that are present are removed.[3] Crabs, carnivorous snails and predatory species of fish should be screened out at the water inlet, and plastic planks are set up along the walls of the pond to prevent their entrance. Filamentous algae, such as *Enteromorpha* and *Chaetomorpha*, are also pests when they grow profusely and affect the fertility of the pond water. Their growth is difficult to control.

The tiny clams planted in November may be harvested the following May or June, when some reach a size of 800–1,000 per kilogramme (Fig 58), and are sold to farmers who grow them for the market. The disparity in size is usually great. The undersized ones are left in the pond for further growth. Usually 50–60% of the tiny clams planted survive and are harvested. Nylon sieves of proper mesh size are used to harvest the clams (Fig 59).

3. GROWING FOR MARKET

3.1 Choice of site

Clams are cultured for the market on sandy flats, tidal estuaries and the inlet and outlet canals of milkfish ponds along the western coast. In recent years, they are also grown in ponds formerly used for milkfish.

When the clams are cultured on sandy flats, the choice of location should take into consideration these points:—

Fig 57 Seed clams sifted to size

Fig 58 Harvesting seed clams

a. Elevation. Elevation with 1–2 hours of exposure at each low tide is preferred. Too long an exposure deprives the clams of much of their food and may cause mortality due to long exposure (over 6 hours) to the hot sun. An elevation with no exposure subjects the clams to destruction by predatory species of fish, crustaceans and molluscs, starfish, etc.
b. Nature of bottom. The bottom should have a sand content of not less than 50%, preferably 60–85%. The clams burrow into soft sandy bottom, and better growth is obtained. On such bottom, the clams also assume more attractive colouration (pink) and have better market value.
c. Salinity of sea water. The sea water should have a specific gravity of 1·015–1·024.
d. Other considerations. Clam beds should be located on bottoms of little wave action. Bottoms with frequent changes of contour and liable to pollution from industrial discharge are not suitable.

3.2 Equipment

Very little physical equipment is required for clam culture on tidal flats. The only installations are bamboo sticks and synthetic fibre nets surrounding the beds to serve as barriers. Clams have a tendency to escape into deeper water with the current. They secrete a mucilaginous thread, which helps them to float in the moving water. Barriers are therefore erected to prevent their escape as well as to screen out predators.

Such barriers consist of an outer fence to prevent the intrusion of predatory fish and other animals and an inner fence to prevent the escape of the clams. The outer fence consists of synthetic fibre nets of larger mesh size (about 5 cm) supported by bamboo sticks reaching above the high water mark (length of bamboo sticks about 2 m). The inner fence consists of nets of smaller mesh size (about 1·5 cm) with a height of about 0·7 m, and its lower edge should be buried under the sand.

The clam growers in central Taiwan simply plant bamboo sticks around the beds at a spacing of 1–2 cm to prevent the escape of the clams. Some just plant bamboo sticks at spacing of 1–3 m to serve as markers with no provision to prevent escape of the clams.

Other equipment includes a watchman's shed built above the beds and a bamboo raft for use in planting, harvesting and other chores.

3.3 Planting

Stocking rate varies a great deal in accordance with the personal

Fig 59 Nylon nets of proper mesh size harvest clam seeds

Fig 60 Harvesting clams on an oyster bed

experience of the farmer as well as his financial status. Generally 2,000–5,000 kg of clam seeds of size 600/kg are planted in a pond of one hectare. Fewer numbers are planted on sandy flats, where as little as 100 kg/ha are sometimes planted.

Planting may be made any time from March to November, in the evening of a cloudy or rainy day, and should be completed just before high tide. This will enable the clams to burrow almost immediately into the sand and so increase their chance of survival. The clams should be planted over the surface as evenly as possible.

3.4 Growth

Growth rate varies with the stocking rate and the environment of the tidal flats or ponds. Generally, clam seeds of 800/kg in size will reach marketable size of 35/kg in 18 months. Best growth rate is obtained from July to September, and there is practically no growth between November and March when the water temperature is low. The following measurements of weight of each individual clam are obtained from clam beds in Tainan having a stocking rate of 20/sq ft.

July 28, 1972 (clams planted)	2·39 g
September 8	5·06 g
October 12	8·20 g
November 17	9·39 g
December 21	9·19 g
January 18, 1973	9·35 g
March 24	9·33 g
May 11	11·66 g
October 5	15·27 g

3.5 Management and care

No fertilizers are applied on tidal flats, and very rarely are the clam ponds fertilized with chicken droppings suspended in baskets on the surface, but the result is not satisfactory. On tidal flats, care should be exercised to prevent poaching and to repair the barriers after a typhoon or rough weather. At low tide, the tidal flats should be inspected, and predatory organisms (crabs, snails, etc.) removed.

Industrial pollution has been the cause of heavy mortality of clams on tidal flats in estuarine waters during the months from March to June.

Clams cultured in outlet canals of milkfish ponds often suffer heavy mortality when water of high salinity and eutrophication is let out from the milkfish ponds.

3.6 Harvesting

The survival rate of clams varies a great deal, as much as 30–75%, but 50% is considered satisfactory. The common gear used for harvesting the clams on tidal flats is the iron rake behind which a bag of proper mesh size to collect the clams is attached (Fig 60). Spades are also used. In pond culture, the clams are collected by hand without draining the ponds.

The clams harvested are put into gunny sacks or plastic bags and transported to the market.

LITERATURE CITED

1. Taiwan Fisheries Bureau: *Taiwan Fisheries Year Book,* 1972.
2. Kuo, Ho: Economic Molluscs of Taiwan, *JCRR Special Publication* No. 38, 1964. (In Chinese.)
3. Huang, T. L.: Aquaculture in Taiwan, *Journal of Bank of Taiwan,* Vol. 25, No. 1, March 1974. (In Chinese.)
4. Lin, Ming-nan: Investigation of Clam Seed Culture. *China Fisheries Monthly,* No. 227. 1971. (In Chinese.)

XI. Culture of Cockle, *Anadara granosa*

1. GENERAL STATUS

The cockle commonly cultured in Taiwan is locally known as blood cockle, *Anadara granosa*. Its meat is red in colour and is a highly priced food served in local restaurants (Fig 61). Although the production of cockles is very large in Malaysia, Thailand and Vietnam, their production in Taiwan is not sufficient to meet the local demand.

The culture area is estimated as around 200 hectares with an annual production of about 120 to 180 metric tons. Most of the production areas are in Chiayi County in southwestern part of the Island. The extent of culture is limited by the supply of seeds.

Another species of less commercial importance is the hairy cockle, *Anadara subcrenata*. Its meat is brownish or purplish in colour and commands a lower price than the blood clam.

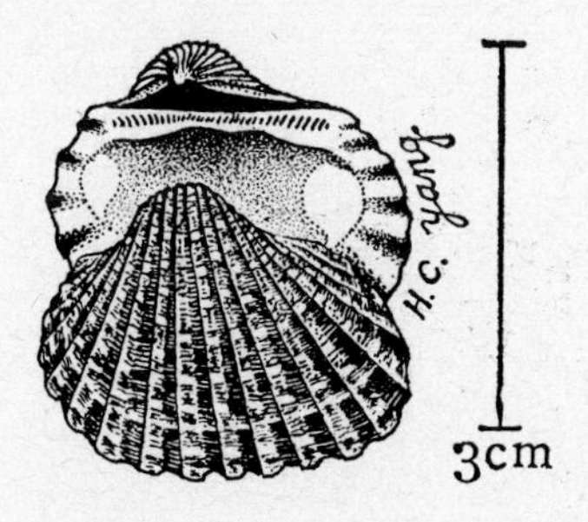

Fig 61 Blood cockle *Anadara granosa*

2. HABITAT AND DISTRIBUTION

In Taiwan, the cockle is found on estuarine mud flats along the western coast of the Island, particularly in the Chiayi and Tainan areas. It generally occurs on shores where there is from 15 to 30 cm of mud and the specific gravity of the water is from 1·008 to 1·014. Elevation with exposure of 5 or 6 hours' duration at low tide is considered suitable.[1]

3. COLLECTION OF SEEDS

The spawning season of the blood cockle in Taiwan is from January to April. The season for collection of seeds is from March to September. They are collected from natural waters mostly off the coast of Chiayi County. Collection is by hand or small-mesh ramie nets and the seeds are placed in baskets or tin cans. Hundreds of people, including women and children, are engaged in this business during the collection season. At the beginning of the season, the seeds collected are of the size of 40,000 to 50,000/kg, and they may reach the size of 800 to 4,000/kg late in the season.

4. COCKLE FARMING

4.1 Selection of site

It is important that the site must not be exposed for more than six hours at low tide. Too long exposure and consequent rise of temperature will cause high mortality of the cockles. Places of continued heavy rain will cause prolonged low salinity and should be avoided. Muddy flats containing up to 90% silt seem most favourable for rapid growth.

4.2 Installations

The cockles remain at one place nearly all the time. Therefore, installations to prevent their escape are not necessary. However, the blood cockle is an expensive item in Taiwan, and installations to prevent theft are necessary. These consist of bamboo sticks or bamboo splittings enclosing the culture area to show that they are private property. A watchman's shed on stilts is sometimes erected on the water.

4.3 Seed farming

Some blood cockle farmers buy or collect the tiny cockle seeds at the beginning of the collection season and grow them to size suitable for stocking. The farms are mud flats usually small in area (0·1 to 0·3 ha) and are enclosed by nylon nettings supported by bamboo sticks. The seeds initially stocked are 20,000 to 30,000/kg in size. They are sown at density of 0·2 to 1·5 kg per mud strip of 2 m in width. In July, the seeds will reach a size of about 5,000/kg and may be sold to farmers who grow them for market.

4.4 Growing for market

Most of the cockle farms are located in Chiayi or Tainan County. The total area is estimated as 200 hectares and is limited by the supply of seeds. Some growers collect their own seeds and plant them at a size of about 70,000/kg in February. Some buy them from collectors in March at the size of 20,000 to 30, 000/kg. Others buy them from seed growers in July, when they reach the size of 5,000/kg. The preferred size is 2,000/kg. Some growers stock clams and/or hairy cockles together with the blood cockles.

As the cockle seeds do not reach stocking size until later in the collection season, the growers usually stock them in June and July. Those stocked at the size of 5,000/kg will reach the size of 500–600/kg in one year, and the marketable size is about 100–200/kg.

There is very little that needs to be done in the management of the cockle beds, except vigilance to prevent theft and possible actions in case of pest and environmental changes. If algae develop profusely, covering the cockles, efforts should be made to remove it. When the beds become too hard due to prolonged use or when there is an increase in sand content of the bottom, growth of the cockles would slow down. They should be redistributed over a wider area, thus reducing the stocking density.

The more common pests are wild ducks, crabs, puffers (globefish) and sea snails. They should be driven away or fenced off.

Environmental changes caused by heavy rain or severe heat are difficult to control. Efforts should be made, however, to prevent damage from industrial pollution.

4.5 Harvesting

Harvesting is by hand. The cockles of marketable size are removed for sale, while the undersized ones are left behind for further growth.

Ordinarily, the cockle seeds stocked at the size of 500–1,000/kg may be harvested by selection after one year of growth and harvested *in toto* after two years. Those stocked at the size of 2,000/kg or larger may be harvested *in toto* after three years of growth.

LITERATURE CITED

1. Kuo, Ho: Economic Molluscs of Taiwan. *JCRR Special Publications* No. 38, 1964.

XII. Culture of the Freshwater Clam, *Corbicula fluminea*

1. INTRODUCTION

Introduced from China, the freshwater clam, *Corbicula* sp. is considered more or less as a nuisance in the ponds and streams of U.S.A. But in Taiwan and some other Asian countries, it is eaten as a delicacy as well as for its medicinal value for treatment of liver disease.

Three species are common in Taiwan, *C. fluminea, C. maxima* and *C. formosana*. The former is the species under cultivation in Taiwan.

The area used for culture of *Corbicula* has not been officially recorded, since it is a comparatively new industry. From an investigation supported by the Joint Commission on Rural Reconstruction in 1973,[1] the total area of *Corbicula* culture (including ponds, reservoirs and drainage canals) is estimated as 1,700 to 2,000 ha, concentrated in the Taoyuan and Changhua areas.

Once very abundant in lakes and streams, a large quantity of *Corbicula* was formerly used as duck feed. Since the fifties, with industrial pollution becoming increasingly serious, they gradually disappeared, and the farmers began to rear them, first in water reservoirs and drainage canals and then in ponds, to meet the local demand.

2. BIOLOGY AND HABITAT

The *C. fluminea* Müller (Fig 62) belongs to the family Corbiculidae. Its sexes are separate, and fertilization is external. The eggs develop into veligers. After the plankton stage, they develop shells and sink to the bottom to lead a sedentary life. They burrow into the bottom and extend their siphons up to the water for respiration and ingestion of food.

In nature, they occur on bottoms of sandy loam where flowing water is

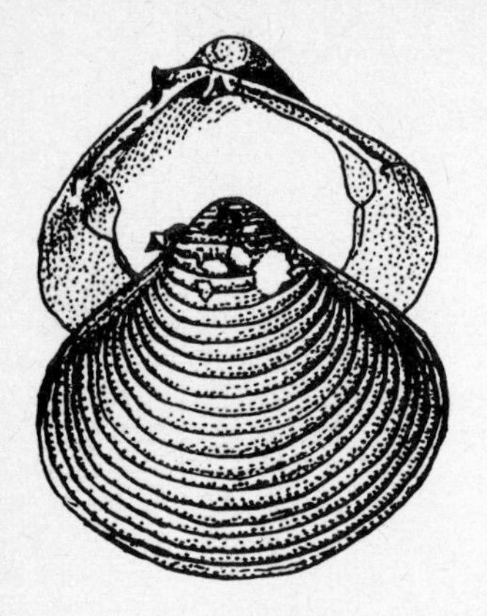

Fig 62 Freshwater clam *Corbicula fluminea*

Fig 63 Drainage canal where *corbicula* are farmed

Fig 64 *Corbicula* of marketable size

present and are found in abundance in the estuarine areas where the water is slightly brackish. They may reach the size of 4·3 × 3·8 cm.

The shell is triangular in shape. The colour of the outer surface of the shell is greenish yellow in the young, but changes with the bottom condition as they grow older, brown on sandy bottom and black on loamy bottom.

3. CULTURE PRACTICE

The *Corbicula* is farmed either in water reservoirs, drainage canals (Fig 63) or freshwater ponds. These water areas must be free from pollution and must not be over-eutrophic. In freshwater ponds, flowing water or aeration should be provided to supply sufficient oxygen to the water. In some of these ponds, baffle dykes are erected to divert the in-coming current to different parts of the ponds. Sandy bottom is preferred. The depth of water is kept at about 1 metre.

The stocking rate is generally 1,000 to 2,000 kg per hectare, varying according to the size of the seed clams and the fertility of the water. The *Corbicula* often reproduce in the cultural areas. So in the second year, less clam seeds are required for planting, especially when a number of under-sized clams are left in the cultural areas.

The seed clams are 800 to 4,000/kg in size. They are purchased from seed collectors or other *Corbicula* farmers, generally in the Changhua and Taoyuan areas. They are packed in gunny sacks or straw bags and transported by trucks. Care should be taken to let them remain wet and not piled too high to reduce mortality on the way.

Before stocking, the ponds are drained and the bottom dried under the sun for two to three weeks,[2] after which the ponds are filled again with water. The farmers then pile the *Corbicula* seeds onto bamboo rafts and scatter them uniformly into the ponds.

No fertilizers are applied when the *Corbicula* are cultured in reservoirs and drainage canals, where they depend on natural foods. But in ponds, soybean meal, wheat and/or rice bran, chicken droppings and inorganic fertilizers (N, P, K) are often applied.

The growth rate of the *Corbicula* varies according to the season, environment and amount of foods available. Generally, planted at a size of 0·11 gm in weight, they reach 0·45 gm after 1·5 months, 0·91 gm after 3 months, 2·25 gm after 4 to 4·5 months, 4·0 gm after 5 to 6 months, and 5·4 gm after 7 to 7·5 months, at which size they may be harvested (Fig 64).

An iron rake with a net attached is used to harvest the *Corbicula*. They are separated by size with a wire sieve, and the undersized ones are returned to the ponds (or reservoirs or drainage canals) for further growth.

4. PESTS

The common carp is the most important predator. So, when fin fish are cultured in *Corbicula* ponds, as is often done, the common carp must be excluded. The fishes commonly cultured with the *Corbicula* are the big head, silver carp and grass carp.

One common disease of the *Corbicula* is blackening of the shell, peeling off of outer covering of the umbo and finally perforation of the shell and death. The cause of this disease has not yet been determined. Possibly it is due to malnutrition.

5. MARKETING

The *Corbicula* are usually sold to dealers who come to size up the ponds, make trial harvest, examine the catch and buy the crop of the entire pond. They are then packed in gunny sacks and transported by trucks to the cities for sale. During the journey, they are kept moist and protected from the sun by a canvas covering.

LITERATURE CITED

1. Fisheries Division, JCRR: Corbicula Culture in Taiwan. *China Fisheries Monthly,* No. 248, 1973. (In Chinese.)
2. Teng, H. T. *et al:* Some Newly Developed Freshwater Cultural Species. *Taiwan Fisheries Research Institute Popular Series* No. 42, 1971. (In Chinese.)

XIII. Shrimp Culture

1. INTRODUCTION

It is not known when shrimp culture began in Taiwan. For centuries, fish farmers in Taiwan have captured juvenile grass shrimps and sand shrimps from the coastal waters and stocked them in milkfish ponds, where they are given no special care and harvested as an extra crop. It is only in recent years, however, that shrimps have been monocultured by some farmers.

Two species of shrimps of commercial importance are cultured in Taiwan, viz. the grass shrimp, *Penaeus monodon,* and the sand shrimp, *Metapenaeus monoceros.*

The grass shrimp, *P. monodon,* and the sand shrimp, *M. monoceros,* are the species commonly stocked in milkfish and *Gracilaria* ponds. According to the Taiwan Fisheries Bureau, the production of grass shrimp from ponds was 140,000 kg and sand shrimp 223,000 kg in 1974.[1] The production was limited by the availability of shrimp seeds. With the production of shrimp seeds by a number of shrimp hatcheries, the production of pond-reared shrimp is expected to increase rapidly in the future.

2. GRASS SHRIMP (Fig 65)

2.1 General characteristics

Also known as sugpo in the Philippines, this is the most important species of shrimp under culture in Taiwan and its culture holds much promise of future development. It is a large shrimp, reaching a maximum individual weight of about 250g, but those reared in ponds seldom exceed 120g.

It is a tropical species, becoming sluggish when the temperature drops below 18°C and succumbing to temperature below 14°C.[2] It is reared, therefore, generally in the southern part of Taiwan where the

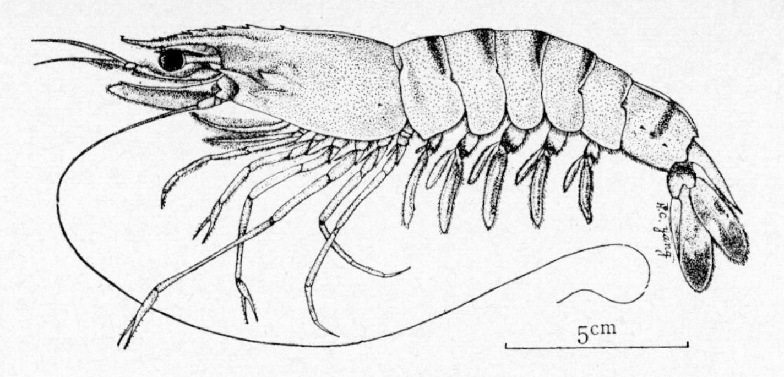

Fig 65 Grass shrimp *Penaeus monodon*

Fig 66 Shrimp hatchery with plastic tanks

winter is mild. It is relatively stenohaline, with a salinity range of 15 ‰ to 35 ‰.

The grass shrimps are caught by commercial trawlers on bottoms of less than 30 m deep in water of low salinity. Unlike many other species of shrimps, they do not hide in the sand, but stay on the surface of the bottom.

2.2 Supply of seeds

2.2.1 Natural supply.

Before the artificial propagation of shrimps was accomplished, all grass shrimp seeds were caught in estuarine waters of low salinity. Since the juveniles like to hide among weeds and grass (hence the name grass shrimp) in shallow water, they are taken by the use of lures, which consist of bundles of grass or twigs set near the shore. The shrimp seeds that gather in the lures are collected with dip nets.

2.2.2 Artificial propagation.

Since 1968, the Taiwan Fisheries Research Institute has successfully accomplished the artificial production of shrimp seeds of *Penacus japonicus, P. monodon, M. monoceros,* and four other species of less importance. Now there are some 30 private shrimp hatcheries producing shrimp seeds, mostly *P. monodon*, with a total production of about 30 million shrimp juveniles.

A shrimp hatchery should be located on the seaside where clean sea water of 25–34 ‰ as well as fresh water (for dilution) is available. The essential equipment consists of first a series of either circular plastic tanks of 0·5 to 1·0 ton capacity and/or concrete tanks of 30 to 70 ton capacity[3] (Fig 66). The number and size of these tanks depend on the scale of the operation. These tanks are used for spawning, hatching and rearing the larvae.

Pumping equipment should be provided to take in sea water. The intake should be placed as far out and as deep as practical and at a point about 30 cm above the sea bottom. The sea water should go through a sand layer or, if necessary, be filtered before it is used.

Aeration equipment is also essential. Air blowers are used for large scale hatcheries, and air compressors are used by small hatcheries. Two sets are usually provided for alternate use. Either an electric generator or diesel engine must be provided to be used in case of power failure.

Fig 67 Dr. I. C. Liao showing grass shrimp spawner

Fig 68 Grass shrimp mysis stage

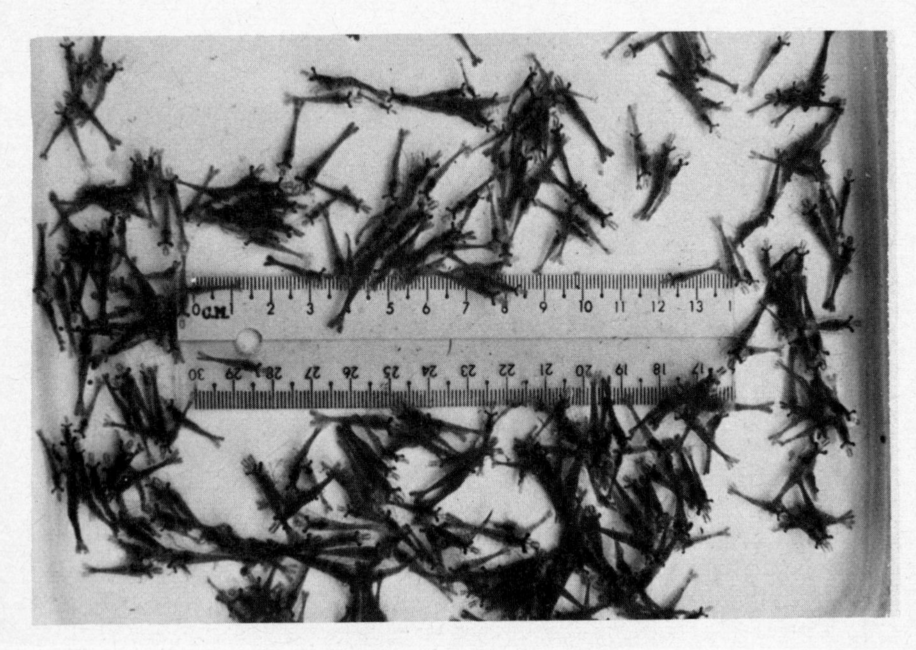

Fig 69 Grass shrimp post larvae

Spawners are obtained from the shrimp fishermen.[4] Until recently, no mature grass shrimps had been found in ponds. It was in May 1975 that the Tungkang Marine Laboratory succeeded in rearing two *P. monodon* to maturity. Eggs were released and hatched, and 2,500 larvae were reared to juveniles. This was accomplished without resorting to removal of the eyestalks (I. C. Liao private communication). The reproduction seasons of the grass shrimp are from March to April and from August to November. The degree of maturity of the spawner is determined by observing the size of the oblong, dark brown ovaries visible through the dorsal part of the shell. The bigger the ovaries, the more mature is the spawner (Fig 67).

One or two spawners are placed in the circular plastic tanks, and more may be placed in the larger concrete tanks. At the beginning of the spawning season, the spawners generally release their eggs in about four days, but at the height of the season, they may spawn on the same night or the next day.

Newly released eggs float on the water surface. They are of roundish shape with a diameter of 0·25–0·27 mm. Each spawner releases about 300,000 eggs. They hatch in about 13 hours at temperatures of 27–29°C and become the nauplius larvae. The eggs and nauplius larvae may be counted by the naked eye in a beakerful of water. If their number reaches 100 per litre, the spawner or spawners should be removed and allowed to continue spawning in another tank.

The nauplius larvae measure 0·28–0·33 mm just after hatching. They do not feed. After about 50 hours at temperatures of 27–29°C, they develop into the zoea larvae. At this stage, they begin to feed and are given small planktons such as *Skeletonema costatum* and *Nitzschia closterium,* marine yeast, fertilized eggs and trochophore larvae of oysters, etc. Skeletonema is the most common feed of zoea larvae. It may be cultured in 0·5 ton tanks. To each ton of sea water are added 100 g of KNO_3, 10 g of Na_2HPO_4, 1 g of K_2SiO_3 (or Na_2SiO_3) and 5 g of $FeCl_3$. Sea water containing *Skeletonema* is then introduced. Under constant aeration and with sufficient light, a brown culture of *Skeletonema* is obtained in 2 or 3 days. The zoëa are very sensitive to light and should be placed away from the light source or dark coloured plastic tanks may be used.

The zoea stage lasts from three to four days, after which they transform into the mysis larvae (Fig 68). At this stage, they resemble the young shrimps, but swim in a vertical position with head down. They feed on the nauplii of Artemia and other zooplanktons such as copepods and rotifers of similar size. Artemia (brine shrimp) eggs are hatched in sea water of 30‰ salinity in about 20 hours at temperature of about 28°C under constant aeration. Rotifers and copepods are gathered from the water supply canals of milkfish ponds with plankton nets.

After 3 or 4 days, the mysis larvae are transformed into post larvae, measuring about 5 mm in length (Fig 69). Up to the 5th day they feed on Artemia or zooplankton supplemented with egg custard (one egg and 3 g of milk powder).[5] After the 5th day they may be fed minced meat of bivalves, trash fish or shrimps or egg custard.[5] From the 5th day on they become benthic, clinging to the bottom or to the walls of the tanks. On the 28th or 29th day, they attain a length of about 20 mm and are suitable for stocking.

2.2.3 Sale and transportation of shrimp seeds

The juveniles or young shrimps are sold by the hatcheries directly to the shrimp farmers. In transportation, they are held in plastic bags, which are filled with sea water and inflated with oxygen. Since the grass shrimp like to stay on the bottom, containers should preferably have large bottom areas.[4] The temperature of the water in the container is maintained at 15° to 18°C by keeping a small bag of ice in the water.

2.3 Over-wintering

Since most of the grass shrimp seeds are caught in the months from August to November, and those from the hatcheries are produced from July to October, it is too late in the season to plant the late juveniles in ponds and get a crop before winter sets in. They are therefore placed in wintering ponds and reared until March or April, when they can be planted in ponds.

Most of the shrimp wintering ponds are in Tungkang, Kaohsiung and Tainan. In Pingtung, the winter is mild, and wintering ponds are simply deeper ponds with no other protection facilities. Stocking is dense and less feeds are given so as to stunt the shrimps. Feeds consisting of cracked snails, small shrimps, trash fish, soybean and peanut meal, etc. are given during warm days (temperature above 18°C). Water is changed weekly. The survival rate is generally 70%. In Kaohsiung and Tainan, the milkfish wintering ponds usually serve the purpose, and oftentimes the shrimps are placed together with the milkfish. The Tainan Fish Culture Station of Taiwan Fisheries

Research Institute experimented on the over-wintering of grass shrimps and found that their survival rate was 28·8 to 49·1% in ordinary milkfish wintering ponds and 68% in a wintering pond specially designed by the Station.[6]

2.4 Pond culture

2.4.1 In association with milkfish

If grass shrimp seeds are in large supply at low cost, probably most milkfish farmers in Taiwan would stock their ponds with some grass shrimp. The cost of grass shrimp juveniles caught from natural water is about US$0·03 to 0·05 each, but the cost of those produced by shrimp hatcheries is only about US$0·02 at the time of writing. So, more milkfish farmers are buying shrimp juveniles to stock their ponds and in larger numbers in recent years. In fact, most of the grass shrimps served in restaurants come from ponds and not from the sea (Fig 71).

In association with milkfish, the number of shrimps planted varies from 5,000 to 8,000 per hectare. They are planted as early as February in the Tungkang area. In Tainan they are planted in early April. They grow to about 40 g each in three months, so at least two crops could be harvested per year. The survival rate is usually above 80%. No special feeds are given to the shrimps, which live off the natural food produced in the milkfish ponds.

The most difficult problem in pond culture of shrimp is the control of predators. The chief predators are *Elops saurus,* gobies, tilapia, jelly fish and the small shrimp *Caridina denticulata* (Fig 70). The last three are most destructive because they multiply fast and their eggs and larvae easily gain entrance through the screen at the water inlet. Predacious fishes can be killed by the application of saponin (in the form of tea seed meal) in concentrations ranging from 2·5 to 10·0 ppm without injury to the shrimps,[7] but the worthless small shrimp *C. denticulata* remains a serious problem both as a predator and a competitor for food. One way to minimize such damages is to keep the young grass shrimps in the nursery pond, where *C. denticulata* rarely occur, until they reach 5 to 8 g in weight before they are planted in the rearing pond.

Another pest is the filamentous green algae and aquatic plants growing abundantly in the milkfish ponds the waters of which are usually less than one metre in depth. Their abundant growth interferes

Fig 70 Worst pest in shrimp culture *Caridina denticulata*

Fig 71 Grass shrimps cropped from milkfish pond

with the movement of the shrimps, and, when they die off, they pollute the bottom of the pond. One remedy is to promote the growth of phytoplankton by minimizing the change of water in the pond so as to retard the growth of the filamentous green algae and other aquatic plants.

2.4.2 Monoculture

Monoculture of shrimp is uncommon in Taiwan. Some monoculture farms are found in the Pingtung area. To avoid as much temperature and salinity changes as possible, water in monoculture ponds is at least one metre in depth. The stocking rate is 30,000 to 50,000 per hectare, and two crops per year are possible. The shrimp reaches 35 g in 6 months, and survival rate is about 50%.

The following is a brief account of an experiment on monoculture of grass shrimp carried out by the Tainan Fish Culture Station in 1974:

(1) *The experimental pond* – A 0·5 ha pond constructed on the tidal land of the Tsengwen area was used. The bottom was sandy, and the walls, 1·8 m in height, were of concrete. Aeration was provided by a blower which delivered air to a 3·5″ plastic pipe on the bottom of the pond. From this main pipe, air was delivered to a number of 1″ plastic pipes which were connected with the main pipe at intervals of 4 m. In these 1″ pipes were many small vents through which bubbles of air escaped to the water surface. A sluice gate was provided with a screen of fine mesh.

(2) *Pre-stocking treatment* – The pond was drained and dried under the sun. To the water puddles that remained tea-seed meal and potassium cyanide were applied to eradicate the pests and predators. After three days, clean sea water was let in to a depth of 50 cm.

(3) *Stocking* – On June 17, 1974, 30,000 post larvae of 30 days after transformation (P_{30}) were planted. They averaged 1·6 cm in length and 0·019 g in weight.

(4) *Feeding* – In the first 10 days, crushed hornshells, *Cerithidea cingulata,* were given as feed. After this, the shrimps were fed exclusively with a specially prepared pellet feed produced by the President Enterprises Corporation. The pellets had a moisture content of 75%. They were given daily in late afternoon. The amounts of feeds given were as follows:

	Kind of feed	Amount given	
1st to 10th day	Crushed hornshells	110% of total weight of shrimp	
11th to 30th day	Pellet shrimp feed	47% of total weight of shrimp	
31st to 60th day	,,	15–21% of total weight of shrimp	
61st to 90th day	,,	10·12%	,,
91st to 120th day	,,	6·52%	,,
After 120th day	,,	5·0%	,,

The pellet shrimp feeds used were made essentially with fish meal, shrimp meal, and plant materials, with the addition of amino acids and vitamins. Its crude protein content was 46·56%.

(5) *Pond management* – Depth of water in the pond was maintained at about 1 m. Water colour was closely watched to maintain a good growth of phytoplankton and control the growth of filamentous green algae by timely change of water. Salinity of water was maintained at from 23·60 to 34·43‰. Water temperature ranged from 18° to 34°C.

(6) *Harvesting* – The shrimps were harvested on November 17th after five months' rearing. The total harvest was 385 kg. Total number of shrimp was 19,439, and survival rate was estimated as 70%. The average individual weight was 20 g, and average length was 14·5 cm.

It is significant that, during harvesting, about 90 kg of *Caridina* shrimps were harvested together with the grass shrimps. This amount was 23·3% of the quantity of grass shrimps harvested and posed a serious problem.

3. SAND SHRIMP

The sand shrimp, *M. monoceros* (Fig 72) are generally cultured in milkfish ponds. It is monocultured only in recent years. It is smaller than the grass shrimp; the marketable size is 5 to 10 g.

3.1 Supply of seeds

Although mass production of sand shrimp seeds by hatcheries is now possible, yet the supply is still from catching in natural water, which is adequate to meet the requirement. They are caught in estuarine waters throughout the island.

3.2 Pond management

Pre-stocking treatment of the pond is the same as in the case of grass

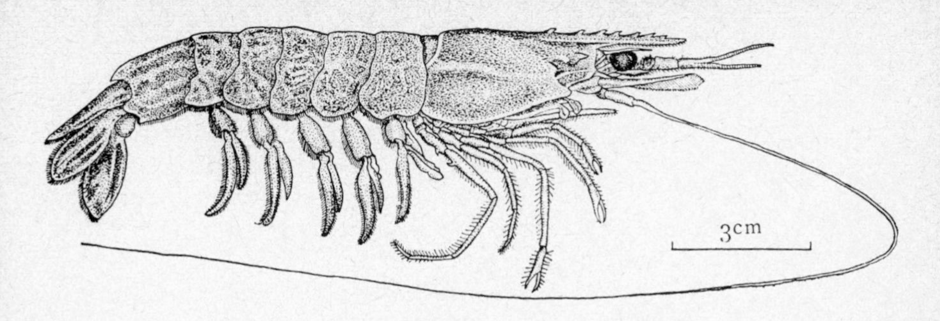

Fig 72 Sand shrimp *Metapenaeus monoceros*

Fig 73 Sand shrimps cropped from milkfish pond

shrimp. Sandy bottom is preferred since the sand shrimp have the habit of burrowing in sand.

When cultured with milkfish, the stocking rate is 60,000 to 100,000 per hectare and no special feeds are given. In monoculture ponds, the stocking rate is over 200,000 per hectare, but may be raised to 300,000 per hectare if the depth of water is over 1·5 metres and aeration equipment is provided. In monoculture, animal protein feeds such as clam meat, trash fish and ground small shrimp should be given; feeds of plant origin such as peanut and soybean meal are also accepted by the shrimps. Other management practices are the same as in culture of grass shrimp.

Suitable water salinity is 15 to 20 ‰, but it is capable of growth even when salinity is as low as 10 ‰.

3.3 Harvesting and marketing

The sand shrimps reach marketable size of about 15 g in about 50 days and may then be harvested (Fig 73). They are generally caught by placing a net at the sluice gate when water is being discharged to take advantage of the habit of the shrimp to swim with the current. They are marketed as medium sized shrimps. As in the case of grass shrimps, they are iced and shipped to the cities by trucks.

LITERATURE CITED

1. Taiwan Fisheries Bureau: *Taiwan Fisheries Yearbook.* 1975.
2. Yang, Hung-chia and Tung-pai Chen: Common Food Fishes of Taiwan. *JCRR Fisheries Series* No. 10. 1971.
3. Huang, Ting-lang: Aquaculture in Taiwan. *Quarterly Journal of Bank of Taiwan,* Vol. 25, No. 1. 1974. (In Chinese.)
4. Liao, I. C. and T. L. Huang: *Experiments on the Propagation and Culture of Prawns in Taiwan.* Paper presented at the Indo-Pacific Fisheries Council Symposium on Coastal Aquaculture. 1970.
5. Lu, T. C.: Experiences in the Mass Production of Grass Shrimp Juveniles. *China Fisheries Monthly* No. 264. 1974.
6. Ting, Y. Y. and T. Y. Huang: Experiment on Over-wintering of Grass Shrimp. *Bulletin of Taiwan Fisheries Research Institute* No. 14. 1968. (In Chinese.)
7. Tang, Yun-an: The Use of Saponin to Control Predacious Fishes in Shrimp Ponds. *Progressive Fish-Culturist,* Vol. 23, No. 1, 1961.

XIV. Crab Culture

1. INTRODUCTION

The crab commonly cultured in Taiwan is the serrated crab, *Scylla serrata* (Fig 74). It is extensively used as food in Southeast Asia, where it is an expensive item of food, relished especially for its ripe gonad.

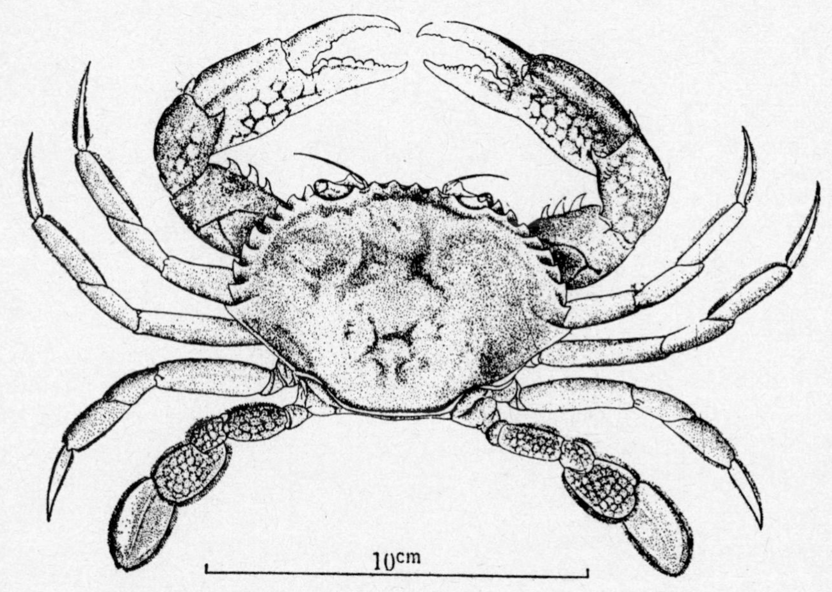

Fig 74 Serrated crab *Scylla serrata*

According to the Taiwan Fisheries Bureau,[1] the production of *S. serrata* in 1973 reached 782 mt, including those captured in coastal waters and those cultured in ponds. There are no data of the area of ponds in which *S. serrata* are cultured as well as their production from ponds. They are extensively stocked in milkfish and *Gracilaria* ponds in southern Taiwan, particularly in Pingtung County. In Ilan County, they are reared in rice paddies together with shrimps and fish after the first rice crop is harvested.

Monoculture of crabs is practised in southern part of Taiwan. The main purpose is to rear the females to maturity, with fully ripe gonad, so as to get a better price.

Three varieties, probably of the same species, are recognized by the fish farmers,[2] viz:

(1) Sand crab. Largest in size; of wild disposition; inhabits flowing water of higher salinity (20–30‰).
(2) Red-legged crab. Smaller in size; shell hard; of wild disposition.
(3) White crab. Medium size; tame and less active; tolerates wider range of salinity.

The white crab is the variety commonly cultured.

Hatching of the eggs of *S. serrata* and rearing the larvae to adults have not been successful in Taiwan. According to Ong Kah Sin of the Malaysian Department of Fisheries,[3] mating of *S. serrata* occurs as early as the first year of life after the female undergoes the precopulatory moult. A male approaching a female in premoulting condition climbs over her, clasps her with his chelipeds and the anterior pair of walking legs, and carries her around. They may remain so paired for 3 to 4 days until the female moults. The male then turns the female over for copulation, which usually lasts 7 to 12 hours.

Sperms are retained by the female, and fertilization may not take place for many weeks or even months after mating. A single copulation may provide sperms for two or more spawnings. The fertilized eggs are attached to the female's pleopods, where they hatch in a few weeks.

The larvae hatch as planktonic zoeae. After passing through several zoea stages and a single megalops stage, which takes about a month, the larvae metamorphose to benthic juvenile crabs.

2. POND CULTURE

2.1 Monoculture

2.1.1 The crab pond

The pond for monoculture of crabs should preferably be located in the estuarine areas where the tidal difference is great to facilitate change of water. The salinity should be from 15 to 30‰. For cleanliness, sandy bottoms are preferred.

The size of the pond is usually small, about 350 m^2. One good arrangement is to divide a large square pond into four smaller square ponds with a concrete tank of about 1·5 metres square in the centre to serve as a water inlet tank for all the four ponds. When water is let in, the crabs will congregate in this tank and can be caught there (Fig 75).

Fig 75 Big square pond divided into four small ponds; concrete water inlet tank at centre

Usually, the walls of the pond, about 1 metre high, are either made of bricks or concrete with protruding lips to prevent escape (Fig 76). Just inside the wall a layer of bricks should be laid on the bottom to form a sloping surface on which feeds can be placed. In ponds with mud walls, bamboo screens are placed obliquely toward the inside of the pond to prevent escape.

2.1.2 Stocking and pond management

The purpose of monoculture of crabs in Taiwan is to rear the females that have just had copulation to the stage of full gonad development. The rearing period is short, one to two months. So increase in weight is not the primary objective. Since only a limited number of post-copulation females can be obtained at one time, a number of small ponds should be provided to hold the different batches of crabs which are collected at intervals.

The seed crabs may be planted any time between April and September. The stocking rate is generally three crabs per m^2, varying in accordance with the water exchange capacity of the pond. The size of the crabs stocked varies from 7 to 12 cm in carapace width.

The feeds usually given are snails, trash fish, fish viscerals, and almost any kind of animal food. Soft-shelled snails (mostly freshwater species) are preferred. If hard-shelled snails are given, they should be cracked first. According to the experience of the crab farmers, snails are the most important feed for the maturation of the crabs and should be given in plenty. The quantity of feeds given daily is about 5% of the total weight of the crabs (about 35 kg per 1,000 crabs). Feeds are given once daily, in the later afternoon since the crabs usually feed after dark. More feeds should be given if the crabs are found to be actively searching for food.

2.1.3 Harvesting and marketing

The crabs are captured by hand with the help of a small dip net or with a large dip net baited with trash fish. Sometimes, plastic or cement pipes are placed on the bottom and lifted up from time to time to empty the crabs that hide therein. Catching them in the water inlet tank as mentioned above seems to be the most efficient method of harvesting. Survival rate is 70 to 90%.

Fig 76 Crab pond with concrete walls

Fig 77 Checking crabs against light to see if ovaries are full

The crabs are generally captured to be sold 20 to 50 days after stocking. Experienced crab farmers are able to tell whether the ovaries are full by examining the crabs against light and also by lightly pressing the shells to feel if they are firm (Fig 77).

During transportation, each individual live crab is bound with a heavy straw rope wetted with water to facilitate handling as well as to keep it moist (Fig 78).

The crabs are sold by number according to their size and not by weight.

Fig 78 Each marketable crab is bound with heavy straw rope

2.2 Polyculture in fishponds

Crabs are also stocked in brackish-water ponds in which shrimp, milkfish and/or *Gracilaria* are cultured. Generally, the ponds are regular fish ponds of 0·5 to 2·0 ha in size, but bamboo or plastic fences should be erected on the mud dykes to prevent escape. Sometimes ponds with concrete dykes are used.

The baby crabs planted are either those of 1·5 to 3·0 cm of carapace width or larger ones of less than 60 g in weight. They are harvested and sold when they attain a carapace width of 12 cm or 220 g in weight.

The number planted varies a great deal, but it usually does not exceed 10,000 per hectare. The purpose of the culture is primarily for growth, so both males and females are planted.

The feeds given are similar to those for monoculture. Adequate quantities must be given to prevent cannibalism.

The crabs generally reach marketable size in five or six months and may be captured for sale. The survival rate is about 50% in those cases where the small baby crabs are stocked and about 70% if larger crabs are stocked. Most of the crabs marketed are male crabs or females without ripe ovaries.

Male crabs that have had many copulations have very little meat and bring a very low price. This is often the case after September. To avoid this, farmers usually harvest their crabs before that time.

LITERATURE CITED

1. Taiwan Fisheries Bureau: *Taiwan Fisheries Yearbook,* 1973.
2. Huang, Ting-lang: Aquaculture in Taiwan. *Quarterly Journal of Bank of Taiwan,* Vol. 25, No. 1, 1974. (In Chinese.)
3. Bardach, John E., John H. Ryther, and William O. McLarney: *Aquaculture,* 1972.

XV. Frog Culture

1. INTRODUCTION

Frogs are a luxury food in Taiwan and many other areas, although they are not eaten in some countries, for instance in India, where frozen frog legs form an important item of export.

The frog commonly eaten in Taiwan is the native green frog, *Rana tigrina* var. *pantherina* Fitzinger (Fig 79). They are small in size and are not cultured.

The species now under cultivation is the American bullfrog, *Rana catesbiana,* which was first introduced into Taiwan by the Japanese in 1924 and again by the Taiwan Fisheries Research Institute in 1951. The first introduction was not successful. The few farmers who reared them soon became discouraged and discontinued the venture. So the frogs disappeared. The tadpoles and young frogs introduced from Japan in 1951 were first reared by the fish culture stations of the Taiwan Fisheries Research Institute. They were propagated and distributed to people who were interested in frog farming.

Because frog farming required very little capital investment and only a small piece of land, many rural people as well as city dwellers became interested in it either as an occupation or avocation. However, nearly all these frog farmers at that time depended on the sales of breeders at an exorbitant price as their source of income. Bullfrogs were not sold as food because of the high cost of production and consumer preference for the native green frogs. So the number of frog farms gradually decreased until recently.

There are at the time of writing some 200 frog farms on the island, most of which are cottage industries. Some larger farms have begun to sell their frogs to restaurants. This is possible by the use of cheaper feeds and improved culture techniques as described in the following.

2. CULTURAL PRACTICE

2.1 Size and construction of ponds[1]

The size of the farms varies a great deal according to the scale of operation. Some make use of only 50 to 60 square metres beside their houses, while the larger frog farms are 5,000 to 6,000 square metres in size.

The ponds usually provided consist of a spawning pond, a hatching pond, a tadpole pond, a pond for young frogs and a pond for growers (larger frogs).

a. *Spawning pond.* The spawning pond is usually 10 to 15 square metres in size. It should be prepared to approximate natural environment with trees and grass planted on the land area to provide shade and shelter. The bottom of the water area should vary in depth, but at least one-third of the water area, which usually comprises one-third of the pond area, should have a depth of 10 cm to serve as the spawning area.

b. *Hatching pond.* The hatching ponds are concrete ponds built in a series to accommodate the different batches of spawns. The water should have a depth of about 40 cm, and a water inlet and outlet should be provided to each pond.

c. *Tadpole pond.* Several ponds with mud bottom and a water depth of 30 to 40 cm should be provided for rearing the tadpoles. To stabilize water temperature and secure better growth, the tadpole ponds should be of a larger size.

d. *Pond for young frogs.* Young frogs are frogs that are less than two months old after transformation. Several ponds should be provided to segregate them according to size. These ponds should have a water depth of 15 to 35 cm, and at least one-fourth of the area should be land which is slightly higher than the water.

e. *Pond for growers.* The ponds for the larger frogs are similar to those for young frogs, but the depth of water should be kept at 30 to 40 cm. In some frog farms, land areas are not provided in these ponds, and floating feed platforms are used instead. Most frog farmers fence their ponds for young and larger frogs with nylon nettings both on the top and sides to prevent escape, but some still use wooden boards or wire screen.

2.2 Selection of breeders

Some frog farmers start their stocks with tadpoles or young frogs, but it is more feasible to start with one or more pairs of breeders as breeding will be done eventually in the farms. The bullfrog starts to breed at the age of about two years. It used to be the belief that an adult frog would only embrace a mate of his own choice,[2] and frog farms sold only what they claimed to be mated pairs. But this has been proved untrue. A mature frog will mate with any competent frog of the opposite sex if given the proper environment.

A male bullfrog can be distinguished from the female by the size of its eardrums, which are twice the size of its eyes; in the female, the eardrums are only slightly larger than the eyes. The male has a bright yellow throat, but the throat of the female is smudgy white, mottled with brown.[3]

The male bullfrog croaks during the breeding season, which is from early March to late August in Taiwan. The sound it makes is similar to the bellowing of a bull, hence its name.

2.3 Mating and spawning

It is customary to select large strong breeders and place them in the spawning pond in the proportion of one male to three or four females and wait for mating to take place. During this time they must not be disturbed.

During mating, the male frog embraces the female in the water from behind with its forelegs pressing the belly of the female. This may last one or two days. Eggs are laid early in the morning and are fertilized in the water. The eggs are surrounded by a jelly-like substance holding them together to form a floating mass on the surface of the water.

2.4 Hatching

The average number of eggs laid by one female is 9,000. Older frogs may lay as many as 12,000 eggs, but the fertilization and hatching rates are low. Usually 1,000 tadpoles can be obtained from each spawning. When an egg mass is observed, it is carefully scooped up in a pan with water and placed gently in the hatching pond. The egg mass must not be inverted while being moved.

The time required for the eggs to hatch varies with the water temperature. At 24° to 27°C, they hatch in about 72 hours. It is desirable to have running water in the hatching pond.

2.5 The tadpole stage

After they hatch, the tadpoles absorb the nutrients stored in their bodies and do not start to feed until the 7th or 8th day, when they may be fed boiled egg yolk. About 20 days after hatching, the tadpoles may be removed to the tadpole ponds.

Proper care and management of the tadpoles is an important part of frog farming, because, the larger the tadpoles at the time of transformation, the larger will be the frogs.

The usual practice is to stock 1,000 tadpoles in one square metre. As many as 2,000 may be stocked if running water is provided. Several small ponds should be provided so that the tadpoles may be segregated according to size to obtain more uniform growth.

The tadpoles are fed fish soluble and peanut meal in the first month, after which they may be fed sweet potato flour, rice bran, kitchen waste and other low-cost feeds. The feeds are placed in a pan near the water surface to facilitate observation. The tadpoles are fed twice daily, once in the morning and once in the afternoon. The tadpoles are omnivorous. They consume both animal and vegetable matter.

2.6 Transformation

The time it takes for the tadpoles to lose (absorb) their tails, grow limbs and transform into young frogs varies a great deal according to temperature, quality of the feeds and the inherent characteristics of the tadpoles, even among tadpoles from the same batch of eggs. Generally, the higher the temperature and amount of feeds, the shorter time it takes for transformation to take place. In central part of Taiwan, the tadpoles from eggs hatched in March transform in about three months, those from eggs hatched in May to July transform in about two and a half months, and those from eggs hatched after August do not transform until after March of the next year and grow to large frogs.

To obtain larger tadpoles, transformation may be delayed by depressing the water temperature (to below 16°C). This is more effective if at the same time sun light is reduced by keeping the tadpoles indoors or under cover. Under-feeding will also delay transformation, but may result in high mortality.

2.7 The young frogs

After transformation, the frogs are moved to the ponds for young frogs. Feeds are not given until all the tadpoles have completed their transformation. They are then segregated as to size, put into different ponds, and feeding may begin. When subsequent observations find some frogs to be significantly different in size, they must be again segregated. This is important to avoid cannibalism.

Over one-fourth of the pond area must be land. When this area is paved with bricks or cement, long exposure of its surface to the sun should be avoided to prevent overheating. When it is earth surfaced, it should be pounded to a hard surface to minimize accumulation of dirt and food particles.

Generally, 100–120 young frogs may be stocked in one square metre of pond, but if they are as large as 6–7 cm in length, the number should be reduced to 60–80.

At the initial stage, the young frogs are fed earthworms, maggots of flies, small fish, small shrimps and other animal foods that are live and moving. The maggots have been proved to be a cheap source of feed for the young frogs. They are produced by placing fish viscerals and trash fish on a wire screen on top of a wooden box. Flies are attracted and lay eggs onto the putrid fish. The eggs develop into maggots, which drop through the wire screen into the box. More food should be given when the temperature is between 20° and 26°C, and the amount of food should be reduced at higher or lower temperature. The average amount of feeds is 10% of the total weight of the frogs, given twice a day. At first, the feed is scattered on the land surface to make it available to all the frogs. Later, it is placed on shallow trays, which are put at the margin of the land (Fig 80). One half of the tray, containing earthworms, maggots, etc. is on land, while the other half containing small fish, shrimp, etc. is partly immersed in water. The purpose of this is to accustom the frogs to a fixed feeding place.

After one to two months, the young frogs are trained to take non-moving dead food. This requires a great deal of labour and patience. Small quantities of dead fish, snails, animal viscerals, etc. are first added and mixed with the live feeds and given to the frogs. The quantity of non-moving food is gradually increased until it completely replaces the live moving food.

In some frog farms, no land area is provided in the pond. Instead, several floating feeding trays are placed on the water surface. As the frogs jump onto the tray, the tray is made to bob and the dead food appears alive. The frogs will then feed on it.

2.8 The growers

When the frogs are over two months old, they are moved into the growers' ponds. For frogs of 12 cm in length, the stocking rate is 50 per

Fig 79 Bullfrog *Rana tigrina* (left female, right male)

Fig 80 Feed trays for young frogs

square metre; for those over 15 cm, the number is reduced to 20–30. The cultural practices are the same as those for the latter stage of young frogs. Meat of the giant African snails is one of the cheapest feeds for adult frogs. The frog farmers buy the snails from the children who gather them and the meat is extracted. Being an agricultural pest, the eradication of this snail is encouraged by the Government.

In semi-tropical Taiwan, the bullfrogs virtually do not hibernate. In cold winter days, they become inactive and cease feeding, but become active again whenever the temperature rises. So they grow at nearly all times of the year. They usually grow to 300 g in seven months after the transformation.

2.9 Enemies

During the tadpole stage, the main enemies are aquatic insects such as dragonfly larvae and diving beetles. Carnivorous fish such as snakehead and catfish also attack and eat the tadpoles. So the ponds should be drained and pesticides applied before stocking.

Birds, snakes and rodents are the chief enemies of the young frogs. Adequate fencing and screening will keep them out.

3. MARKETING AND TRANSPORTATION

Taiwan has not yet developed an export trade of bullfrogs. All the cultured frogs are sold by the farmers directly to restaurants. The common market size is 150–220 gm which the frogs reach in 8 to 10 months. At this size, they bring the best price.

For short distance transportation, they are simply held in gunny sacks or any cloth bags, but for long distance shipping they are held in wooden boxes. They are kept cool and wet during the journey.

LITERATURE CITED

1. Lin, Chiu-chang: Frog Culture. *Taiwan Fisheries Bureau Fish Culture Pamphlet* No. 49, 1973. (In Chinese.)
2. Wright, H. H.: Frogs. Their Natural History and Utilization. *Report U. S. Bureau of Fisheries,* 1919.
3. *Florida Department of Agriculture:* Bullfrog Farming and Frogging in Florida. *Bulletin* No. 56, 1952.

XVI. Culture of Soft-Shell Turtle

Although farming of the soft-shell turtle, *Trionyx sinensis* Wiegmann (Fig 81), had its beginning when Taiwan was under Japanese occupation, the scale of operation at that time was very small. People were not interested in turtle farming, because the animal takes a long time, two years or more, to reach marketable size. The first turtle farm of larger size was started in Taichung in 1926. The farm had only four small ponds, totalling 155 sq m.

Attracted by the high price paid by restaurants for the soft-shell turtle in recent years, about US$6/kg at its highest, more and more people have taken to turtle farming, and the scale of operation has become larger. According to the Taiwan Fisheries Year Book, the total acreage in turtle culture was 104 ha with a production of 259 mt in 1972.

1. GENERAL DESCRIPTION OF THE SOFT-SHELL TURTLE

The soft-shell turtle actually has a poorly ossified hard shell, but it is soft-edged and lacks horny scales. The male can be distinguished from the female by the following characteristics:

1. The tail of the female is shorter than that of the male, not protruding outside the carapace.
2. The carapace of the female is less oval than that of the male.
3. The body of the female is thicker than that of the male.
4. The distance between the two hind legs is greater than that of the male.
5. Mature males are much larger than the females, often twice in size.

It inhabits freshwater and may be found in all the streams, lakes and ponds where the bottoms are muddy. Respiration is by lungs. It loves to climb water edges or aboard any floating object to get the warmth of the sun.

It is extremely timid, disappearing into the water at the approach of

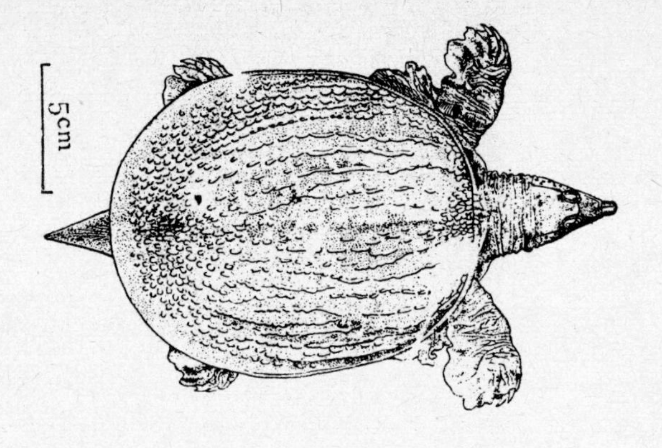

Fig 81 Soft-shell turtle *Trionyx sinensis*

Fig 82 Turtle ponds

humans or other animals. But it is also viscious and cannibalistic, biting the hand that comes close to it.

It is active and feeds well during the warm weather. When the water temperature drops below 12°C, it buries itself in the mud to hibernate.

2. POND CONSTRUCTION AND WATER SUPPLY

Ponds are of mud bottom. The sides may be of mud, brick or concrete. The walls are provided with protruding lips on the top to prevent the turtles from escaping. The lip extends about 12 cm beyond the inside of the walls.

The ponds should be located in an open area with good sunlight (Fig 82). The water should be from a warm source of supply. Cold spring water is not suitable. Still water with rich growth of phytoplankton is preferred. It not only gives the turtles a sense of security, but the process of photosynthesis also adds oxygen to the water and removes obnoxious gases formed by decomposition.

Since the turtles are cannibalistic, several ponds should be provided so that they may be segregated by size. The ponds vary from 700 to 1,700 m^2 in size with a water depth of 40 to 80 cm and a freeboard of at least 30 cm. The deep portion of the pond should have 15 to 30 cm of soft mud where the turtles can hibernate in the winter. The bottom of the shallow portion should preferably be a mixture of sand and mud. The larger turtles have a tendency to escape at the corners of the pond, so the lip at these locations should extend further out, or a triangular piece of wooden board could be added to each corner. As in all fish ponds, the bottom should slope toward the water outlet to facilitate drainage.

Besides the pond for growers as described above, some turtle farms also have ponds for spawners. These are generally 70 to 1,000 m^2 in area and should be located in the quieter part of the farm, free from disturbance. In larger farms a small brick or wooden house, in the form of a kennel or a lean-to, is constructed at one corner of the pond for the turtles to lay their eggs (Fig 83).

The smaller farms generally have no special pond for the spawners, and a small house or a platform with sandy bottom for egg laying is provided at one corner of the rearing pond (Fig 84).

It is necessary to provide nursery ponds for the young turtles to prevent cannibalism by the large turtles. They are usually located close to the house of the farmer or caretaker and vary in size from 10 to 600 $m.^2$ The bottoms are paved with sand, and a top cover is usually provided.

3. CULTURAL PRACTICES

3.1 Rearing in nursery pond

The farmers start operation by buying young turtles from other turtle farmers or dealers. The young or seed turtles are 2–3 cm in length. They start feeding almost immediately. In the nursery pond, they are fed tubifex worms and crushed shrimp. Two feedings are given each day. The amount of feed given daily is approximately equal to about 10% of the total weight of the turtles. Later, minced fish and boiled yolk of egg in the daily amount of 20% of the total weight of the turtles may be given. Mixed feed for eel is also generally used in the daily amount of 5% of the total weight.

Mortality is generally high in the nursery ponds, especially in the second and third months after hatching. This is due to accumulation of unused feed and detritus on the bottom and consequently decomposition with the production of methane, hydrogen sulfide, ammonia, etc. Regulation of feeding and timely change of water are necessary to reduce loss.

To maintain a higher temperature for growth, some nursery ponds are covered by transparent plastic sheets stretched on a wooden frame. The sheet should be removed if the water temperature exceeds 32°C.

The turtles in the nursery pond will reach 200 g in weight in a year and may be moved to the rearing or growers' pond.

3.2 Rearing for market

In the growers' pond, the turtles are segregated by size and put into different ponds to minimize cannibalism. They are fed mainly trash fish, which is often mixed with corn or soybean meal. Low-grade formulated eel feed is also used in the daily amount of 5% of the total weight of the turtles. Feeding is once to thrice daily according to observation of the appetite of the turtles and the water temperature. Fatty fish may be used as feed in the winter months, but lean fish should be used in the summer. Salted fish should never be given, as this is often the cause of high mortality.

The feeds are placed either on a wooden tray or a bamboo basket and lowered below the level of the water.

In southern Taiwan, the turtles feed almost year round, but in central and northern Taiwan they cease to feed from December to February when the temperature is below 15°C. Feeding is best when the water temperature is between 20° and 26°C.

The rate of growth depends a great deal on the climate. In central and northern Taiwan, the turtles will reach the market size of about 600 g in two and a half to three years, but in southern Taiwan, where they feed almost the year round, this size is reached in one and a half to two years. In the Chiaochi hot-spring area, with the use of geothermal water, it is reported that they reach 600 g in one year.

For turtles of one to three years in age, the feed conversion rate is 6–8 using trash fish and 2 using dry mixed feed (for eel). From the fourth year on, the feed conversion rate is lower and growth slower. This may be due to the fact that the mature turtle spends a great deal of energy for reproduction.

The stocking rate varies with the age of the turtles as shown in the following table.

Age	Number per ·01 ha
1-year	1,500–3,000
2-year	600–1,800
3-year	300– 450
4-year and over	90– 300

Some farmers put a small number of walking catfish, *Clarius fuscus*, in the turtle ponds to take care of the feed not consumed by the turtles.

When the turtles reach market size, they are captured by nets dragged across the bottom or by hand. When the ponds are drained of water, the farmers wade into the ponds and grab the turtles from the rear. Care must be taken to avoid being bitten by the turtle. If bitten, placing the turtle back into the water will cause it to release its hold.

4. SPAWNING

The spawners are either reared in special spawners' ponds or in the rearing ponds in which the three-year-olds are kept. In some farms, a small brick or wooden house, usually 10–20 m^2 in area and 100 cm high, is built. It is provided with a small window and a door. This is for spawning and incubation of the eggs. The floor of this house has a gentle slope and is covered with clean sand to a depth of about 15 cm. One or two pieces of wooden planks are placed at the small door (or hole) of the house sloping with the end into the water to serve as a ladder for the turtles to climb in to lay their eggs in the sand. In smaller farms, an elongated sloping platform with sand on the bottom is provided at one corner of the pond for the turtles to lay

Fig 83 Small brick house for egg-laying

Fig 84 Platform for turtle egg-laying

their eggs on. At the lowest part of the floor of the small house or platform, a small basin containing water is placed level with the topmost sandy floor. The young turtles will instinctively drop into the basin after they hatch (Fig 85).

The spawners should be at least three years old. Although turtles in southern Taiwan start spawning at two years of age, their eggs are few in number, irregular in size (usually smaller) and do not develop into healthy young turtles. It is best, therefore, to use as spawners large turtles eight to nine years old. One or two males to one female is usually considered appropriate.

The turtle makes a hole in the sand and lays its eggs therein (Fig 86). The turtle farmer removes the eggs and examines them to see if they have been fertilized. The eggs are spherical, 1·5 to 3·0 cm in diameter. They are grayish brown in colour. Unfertilized eggs develop large white spots in a few days. The good eggs are arranged in rows with the air cell up and spaced about 2 cm apart and covered with about 5 cm of sand to incubate. They are usually marked with numbers and dates. The sand should be kept warm and slightly moist. At average temperature of 30°C, the eggs hatch in about 50 days. Hatching rate is higher in an incubation house than on an open platform, often exceeding 80%.

5. PREDATORS AND DISEASES

The most important predator is the snake, which devours the turtle eggs and young.

The highest mortality occurs when the young turtles are three months old, when they are infected with mould. This is the result of high-density stocking and consequent injuries from biting among the turtles themselves. Density should be reduced and the infection treated with 100 ppm formalin or 2 ppm malachite green.

The most prevalent and most serious disease is what is known as swelling of the neck. It begins with the appearance of red spots on the plastron and sometimes blindness, followed by swelling of the neck, carapace and plastron. The disease is highly infectious and often causes mass mortality. The common treatment of giving the diseased turtles antibiotics and sulpha drugs is not very effective.

Fig 85 Small water basin on spawning platform for baby turtles to drop into

Fig 86 Turtle eggs

XVII. Culture of *Gracilaria*

1. INTRODUCTION

Four species of *Gracilaria* of economic importance are recognized in Taiwan: *G. gigas* (or *G. chorda*), *G. confervoides, G. lichenoides* and *G. compressa.* Of these the *G. confervoides* is the species most commonly cultured in ponds.[1] *G. gigas* is also cultured in some areas.

Gracilaria and *Gelidium* are the most important raw materials for the manufacture of agar. Culture of *Gracilaria* (Fig 87) started in 1962, when demonstration in the Hsinta Lagoon (Kaohsiung) was started by the Chilou Fishermen's Association. Later in that year, an experiment on the culture of *Gracilaria* by attaching them to ropes was carried out in Tainan with the assistance of the Joint Commission on Rural Reconstruction. Since then the culture of *Gracilaria* in ponds formerly used for milkfish has gradually gained popularity. According to the Taiwan Fisheries Year Book, the total culture area in 1973 was only 112 ha, but the area in 1974 was estimated as 400 ha. Pingtung County alone accounted for 110 ha, with a production of 1,000 mt of dried seaweed.

2. REPRODUCTION

Gracilaria is a genus of red algae, with about 100 species, and is widely distributed over most parts of the world. It has a cylindrical or flattened thallus, which is usually bushy and attached to the substratum by a small discoid base. The thallus is dichotomously, irregularly or proliferously branched.

Gametophytes of *Gracilaria* are heterothallic. The male thalli produce spermatangia. The eggs produced by the female gametophyte are fertilized and develop into carposporophytes, which produce carpospores. The carpospores develop into tetrasporophytes. Tetraspores are produced from tetrasporophytes and germinate into male and female gametophytes.

3. CULTURE IN PONDS

3.1 The pond

In nature, *Gracilaria* grow best in water of comparatively low salinity and low wave action. Inlets and shallow sea with a salinity of 8 to 25‰ (specific gravity 1·005–1·020) are suitable for their growth. The best growth is obtained at temperatures of 20° to 25°C. In the pond, they begin to die when the salinity exceeds 35‰ (specific gravity 1·024).[2]

Most of the *Gracilaria* ponds in Taiwan are located in southern part of the island where high temperature prevails. In the selection of pond site, the following should be taken into consideration:

a. The area should not be exposed to strong wind, which will pile up the *Gracilaria* on one side or one corner of the pond.

b. Freshwater should be available to dilute the water when salinity becomes too high due to evaporation.

c. There should be sufficient tidal difference to facilitate change of water.

d. The bottom should be of sandy loam.

e. The water should have a pH value of 6 to 9, preferably 8·2–8·7.

Most of the ponds now in use are those formerly used for milkfish. They are small in size (about one hectare) and rectangular in shape, with the long axis perpendicular to the direction of the wind. Preferably a windbreak should be erected on the windward side.

3.2 Planting stock

Due to its adaptability to a wide range of environmental conditions, its high cropping rate and high jelly strength of the dried product, the *G. confervoides* is the most extensively cultured species in Taiwan.

The cuttings or torn sections of the *Gracilaria* are usually used as planting stock. The *Gracilaria* farmers buy the planting stocks from other farmers. Each plant is either torn apart at the base by hand or cut into pieces, which are planted uniformly on the bottom of the pond.

It is important to plant only healthy stock. The healthy plants are those which (1) feel elastic on touch, (2) possess many shoots or stems with tips of reddish brown colour,(3) have heavy or thick stems, (4) feel brittle on being bitten, (5) have straight stems with straight ends, and (6) are free from adhesion of detritus and other foreign materials.

The planting stocks are transported by trucks early in the day or in the cool of the evening. They are frequently sprinkled with sea water on the way and covered with wet straw or gauze to minimize evaporation.

Perforated pipes of plastic or bamboo are inserted into the bottom portion for aeration to prevent over-heating. Upon arrival at the destination, the planting stocks are immediately put into the pond.

3.3 Pond management

From 3,000 to 5,000 kg of the fragmented plants are planted in a pond of one hectare in size. Too low a stocking rate will result in profuse growth of phytoplankton. Planting time is usually in April. The planting stocks are placed uniformly on the bottom. They are usually fixed on bamboo sticks planted on the bottom or covered with old fish nets to prevent them from drifting to one side or one corner of the pond.

The depth of pond water is generally 20 to 30 cm in the months from April to June. With the increase of temperature after June, it is increased to 60 to 80 cm. One change of water in two or three days is the general practice. The frequent change of water is necessary to maintain the proper salinity as well as to provide the nutrients for the growth of the *Gracilaria*. During the dry season, some water should be introduced every day to avoid undue increase of salinity due to evaporation. During rainy days, the pond water should not be changed, sometimes for as long as a week, but if the salinity of the pond water drops too low, introduction of water of high salinity would be necessary.

To accelerate the growth of the *Gracilaria,* fertilization of the pond water is necessary. Either organic or inorganic fertilizer is used. In some areas three kilogrammes of urea are applied weekly to a one-hectare pond; while in some farms, 120 to 180 kg of fermented manure from pigsties are applied to each hectare of the pond every two or three days at the time when water is being introduced.

Since *Gracilaria* do not tolerate temperatures below 8°C, the depth of the pond water should be increased when winter sets in, or they may be moved into wintering ponds, which are deeper and protected with windbreaks.

The main pests of the *Gracilaria* are the filamentous algae, which develop profusely, especially in the winter months. They envelop the *Gracilaria,* depriving them of nutrients and light, as well as making them difficult to clean when being harvested. Such algae are mainly *Acanthophora, Bangia, Hormisia, Entermorpha* and *Chaetomorpha.* The last two cause the most harm and are most difficult to control. The common practice for control of the algae is to lower the water level and decrease the turnover of the pond water for 7 to 10 days. At the same time, more *Gracilaria* plants of larger size and more vigorous growth are introduced to absorb the nutrients which otherwise would be taken up by the pest algae. About 500 to 1,000 Milkfish per hectare (size of 150 g or more) may also be introduced at the early stage to control the green algae (*Enteromorpha* and *Chaetomorpha*). They will consume the *Gracilaria* after the green algae are gone. When this happens, efforts should be made to remove them by netting them at the water inlet where they congregate. *Acanthophora* look a great deal like *Gracilaria* and are removed by hand at harvest.

Many *Gracilaria* farms stock their ponds with grass shrimps, *Penaeus monodon*, or crabs, *Scylla serrata*, to obtain additional income.

3.4 Harvesting, drying and packaging

The *Gracilaria* are harvested once every ten days from June to December. Harvesting is by hand or by the use of scoop nets. They are uprooted and washed in the pond water to remove the mud, detritus, sand and snails that adhere to them. After being thus cleaned, the *Gracilaria* are spread uniformly on bamboo screens or plastic sheets (usually placed on the sides of the ponds) to be dried under the sun. Sometimes they are spread on a concrete surface to dry. The drying ratio is about 1:7. Generally 10,000 to 12,000 kg of dried *Gracilaria* are produced from one hectare of pond annually.

According to standards set up by the Bureau of Standards in Taiwan, dried *Gracilaria* for export should contain not more than 1% of mud and sand, not more than 1% of mollusc shells and not more than 18% of other weeds. In other words, foreign materials should not exceed 20%. Moisture must not exceed 20%.

The dried *Gracilaria* are packed in gunny sacks in quantity of 100 kg each, either for export to Japan or sold to local agar plants.

LITERATURE CITED

1. Cheng, C. H., Ming-nan Lin and Tung-liang Lee: Culture of *Gracilaria*. *Taiwan Fisheries Bureau Technical Series*, 1968.
2. Workshop on *Gracilaria* Culture. *Taiwan Fisheries Bureau*, 1968.

Fig 87 Experimental culture of *Gracilaria* on plastic ropes

XVIII. Other Cultured Species

In addition to the cultured species described in the previous chapters, the following species are being cultured but have not attained commercial importance:

1. RAINBOW TROUT

The rainbow trout, *Salmo gairdneri* (Fig 88), was first introduced from Japan in 1957. Three thousand fertilized eggs were brought in, hatched and liberated into a reservoir, where they probably perished. In 1959 and 1960, 200,000 eyed eggs were introduced each year. They perished due to lack of facilities and typhoon damage respectively. In 1961, 50,000 eyed eggs were imported and were hatched and reared to adults in a newly established hatchery at Ma-Lun located high on the upper stream of Tachia River.

The Ma-Lun Trout Hatchery has been in operation since that time (Fig 89). Each year it produces 20,000 to 30,000 fingerlings. In 1974, the Hatchery sold 3,450 trout with a total weight of 1,623 kg, and 10,000 fingerlings were liberated into the Wanta Reservoir.

The feeds for the fingerlings consist of animal viscera. For larger fish, a mixture of animal protein food (55%) and plant protein food (45%) is fed.

In addition to the Ma-Lun Trout Hatchery operated by the Taiwan Fisheries Research Institute, a private trout hatchery located at Kinshan (near Taipei) is now producing 5 to 7 mt of trout each year, most of which are sold to tourist hotels.

Due to the difficulty of finding sites with adequate supply of cold water, there is very little possibility of expanding trout farming in Taiwan.

It may be mentioned in this connection that a land-locked salmon, *Oncorhynchus masou* (Brevoort) is indigenous in the mountain streams of Taiwan. Due to indiscriminate fishing by the aborigines, it has now practically disappeared.

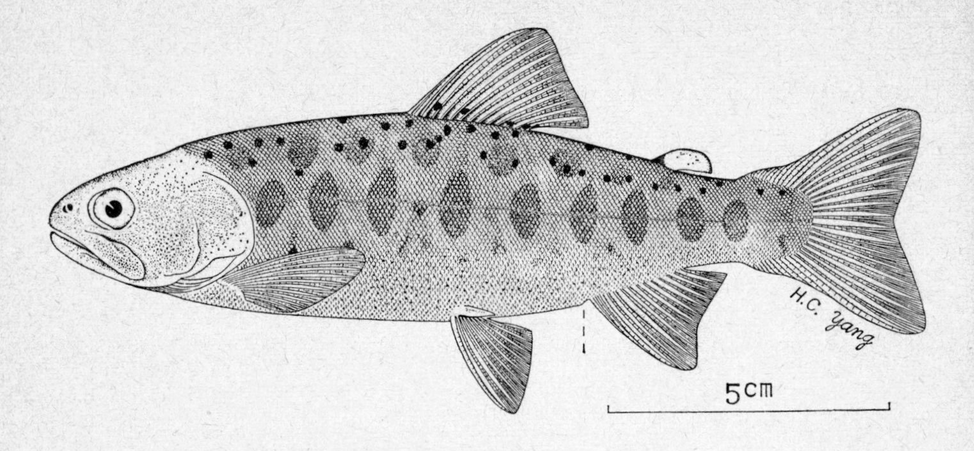

Fig 88 Rainbow trout *Salmo gairdneri*

Fig 89 Trout ponds at Ma-lun

2. AYU

Ayu, *Plecoglossus altivelis* (Fig 90), is a smelt-like fish. It is the only anadromous fish known in Taiwan and used to enter the streams of Taipei, Taoyuan and Miaoli, but is now practically extinct due to the use of dynamite and the lack of protection.

The ayu is a highly esteemed table delicacy, especially in Japan. There are two ayu farms in northern Taiwan. They import fingerlings from Japan and rear them in freshwater ponds. They are sold to local Japanese restaurants at a size of about 20 cm.

The culture technique is similar to that of trout culture. Differing from trout, they tolerate higher temperature and lower dissolved oxygen content.

Efforts to rear the hatchlings (Fig 91) to stocking size have not been successful, although the ayu could be spawned in Taiwan.

3. CATFISHES

Besides the native walking catfish, *Clarius fuscus,* three other species of catfish are known to have been reared in ponds in Taiwan. They are (1) the Japanese common catfish, *Parasilurus asotus,* (2) Thai catfish, *Pangasius sutchi,* and (3) the walking catish from Thailand, *Clarius betrachus*. In addition, some fish farmers have imported fingerlings of the channel catfish, *Ictalurus punctatus,* and rear them on trial.

3.1 The *P. asotus* (Fig 92) is a large freshwater catfish found in the western part of Taiwan, reaching 30 cm in length. Its food consists of molluscs, frogs and small fish. It spawns in the months from April to June. The eggs adhere to aquatic plants. The hatchlings inhabit bottoms of dense vegetation growth. They grow to 10–15 cm in a year and mature in 2 years.

The rearing ponds are 350 to 700 m^2 in size and about 1·0 to 1·5 m in depth. A covered feeding platform should be provided, because it prefers to feed in the dark.

Fingerlings are purchased from fish fry dealers. About 500 g of fingerlings in weight are planted in 3·5 m^2 of water area. Feeds, consisting of minced trash fish mixed with formulated eel feed, are placed in a wire basket and lowered just beneath the water surface. The quantity of feeds given daily is about 10% of the weight of the fish, varying according to temperature and other environmental factors.

Cultured in ponds, the *P. asotus* reach 200–300 g in weight by the end

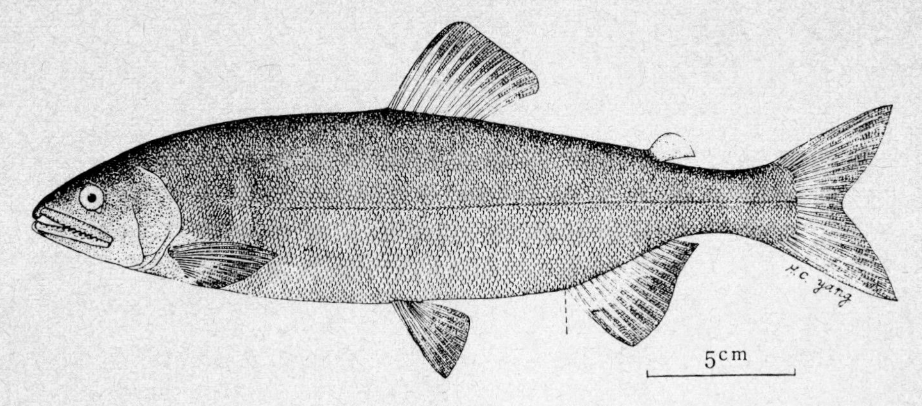

Fig 90 Ayu *Plecoglossus altivelis*

Fig 91 Stripping the ayu for eggs

Fig 92 Japanese common catfish *Parasilurus asotus*

of the year and may be harvested. They are impounded in clean water and marketed the next day. They are sold alive.

In 1971 and 1972, the Chupei Fish Culture Station of the Taiwan Fisheries Research Institute succeeded in an experiment on the artificial propagation of the *P. asotus*.[1] After 30 days, the hatchlings assumed the appearance of the adult fish. *Brachionus* and daphnia were used as feeds for the larval fish.

3.2 The Pangasius catfish was introduced from Thailand through the effort of S. W. Ling in 1969. Only a few live specimens now remain in the fish culture stations, and they are as yet immature at time of writing. The difficulty with this species is that is cannot tolerate the low temperature of winter in Taiwan.

3.3 *C. betrachus* found their way into Taiwan through importation by private fish farmers. They grow faster and to a larger size than the indigenous *C. fuscus* and are gaining popularity. Fingerlings are now produced locally by artificial propagation.

4. SWAMP EEL

The swamp eel, *Fluta alba* (Fig 93), is served as a delicacy in restaurants and commands a high price. It is caught mainly from rice paddies and has not been cultured, except that some are impounded to meet the off-season market demand.

In 1969–1970, the Chupei Fish Culture Station of Taiwan Fisheries Research Institute carried out an experiment on the culture of the swamp eel. Three cement tanks with mud on the bottom were used. The feeds were earthworms, minced trash fish and eel feeds.[2] The conclusions reached were:

(1) The swamp eel could grow by artificial feeding, but the feed conversion ratio was very high, 19·5 to 25·2.
(2) They could be trained to feed in day time.
(3) Their feeding activity slowed down when water temperature was down to 16°C and ceased entirely at 12°C or 13°C.
(4) The amount of feed given was 3 to 5% of the total body weight.

5. POND LOACH

The pond loach *Misgurnus anguillicaudatus* (Fig 94), is a popular food fish in Japan and is served in Japanese restaurants in Taiwan. It is also consumed by the rural population in Taiwan. It is a warm-water fish, with an optimum water temperature of about 25°C. When water temperature drops below 8°C or exceeds 32°C, it burrows into the mud bottom and becomes sluggish. It is generally found in rice paddies, mud ponds and other stagnant waters. It is capable of breathing through the skin and the intestines and can live in water of very low dissolved oxygen content. The spawning season of the pond loach is April to August, with the peak from the end of May to the end of June. The eggs adhere to aquatic plants and hatch in 2–3 days.

Only a few fish farmers in the Taiwan area are engaged in the culture of the pond loach. The ponds are similar to the eel ponds, but 20–30 cm of mud should be added to the bottoms. The fry are collected from the wild. They are 3–5 mm in length when newly hatched. After 2–3 days, the yolk sacs disappear, and they may be fed with egg yolk and milk powder. A little later, natural feeds such as daphnia and rotifers, supplemented with fish meal, rice bran and other artificial feeds, may be given. In about a year, they exceed 10 cm in length and may be marketed.

The stocking rate is about 130 g of fry to 3·5 m^2 in the nursery pond. They are transferred to the rearing pond (0·01–0·02 ha in size) when they reach 5–7 cm in length at density of 300–500 g per 3·5 m^2.

Often they are reared in ponds converted from rice paddies, in which case the ponds should be fenced with plastic sheets to prevent their escape.

They are harvested by draining the ponds and should be impounded in clean water for one or two days to empty the guts and remove the muddy taste before marketing. They are sold alive.

6. KURUMA SHRIMP

The kuruma shrimp, *Penaeus japonicus* (Fig 95), is the favourite shrimp of the Japanese people and is the species under extensive cultivation in Japan. Small quantities are airlifted from Taiwan to Japan as live shrimps, which bring very high return.

In Taiwan, its culture is limited to a few farms on the Penghu Island (Pescadores) due to the following reasons:

(1) The tidal difference along the coasts of Taiwan is too small to allow the construction of ponds with a depth of water over one metre unless pumping facilities are installed. Deeper ponds are required by the kuruma shrimp to minimize changes in water temperature and salinity.

(2) The large quantity of eutrophic water discharged from milkfish ponds makes it difficult to obtain from the canals and estuaries the clean sea water required by the kuruma shrimp.

(3) Because of unevenly distributed rainfall, it is difficult to control the salinity of the pond water.

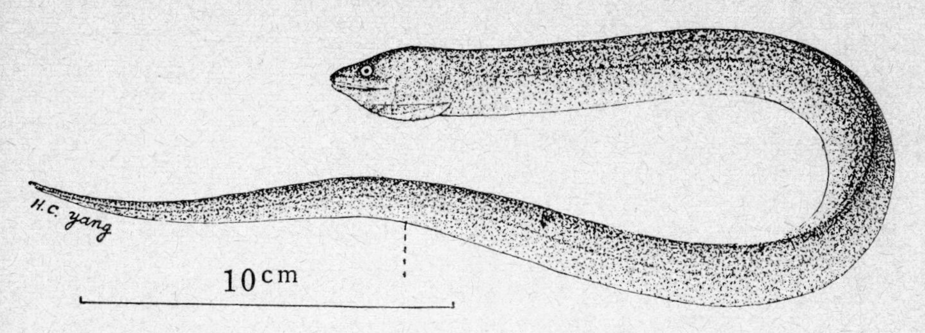

Fig 93 Swamp eel *Fluta alba*

Fig 94 Pond loach *Misgurnus anguillicaudatus*

Fig 95 Kuruma shrimp *Penaeus japonicus*

Fig 96 *Laternula* sp, clam commonly used as feed for shrimp

(4) Most of the ponds in Taiwan have bottoms of sandy loam, which are difficult for the kuruma shrimp to burrow in. Nor is it practical to add sand to the bottom, because the sand found on the coast of Taiwan is too fine and forms a hard surface in water.

(5) The best feeds (Fig 96) for the kuruma shrimp are clams, small shrimps and squids, which are expensive in Taiwan.

Production of kuruma shrimp seeds is no problem. Plenty of spawners are available, since many shrimp boats are equipped with facilities to keep the kuruma shrimps alive for export. But practically no kuruma shrimp seeds are produced commercially, because the demand is small.

The kuruma shrimp farms on the Penghu Islands follow more or less the method for monoculture of grass shrimp with little modifications and are generally not very successful.

7. GIANT FRESHWATER PRAWN

The giant freshwater prawn, *Macrobrachium rosenbergii* (Fig 97), was introduced from Thailand in 1970, through the effort of Dr. Shaowen Ling, FAO Inland Fishery Biologist. From one male and one female survivors from this shipment, hundreds of thousands of offspring have been bred by the fisheries laboratories and commercial hatcheries (Fig 98). At present, there are several shrimp farms engaged in rearing the *Macrobrachium* prawns, but none of these prawns have yet reached marketable size. So, the fate of this enterprise, which depends on price and cost of production, is yet undetermined.

The giant freshwater prawn is widely distributed in most of the tropical and subtropical areas of the Indo-Pacific region, occurring in both fresh and brackish waters,[3] but only a few wild specimen have been caught in southern Taiwan. It is omnivorous. Growth rate is rather rapid, reaching a maximum of 125 g in weight in about six months.

The facilities and equipment for hatching and larvae rearing are similar to those for the grass shrimp. Since the new hatchlings require a brackish water medium, the hatchery should preferably be located near the sea.

Berried females are kept separately, one in each tank of 0·5 to 1·0 ton capacity.[4] The fresh water used should be kept clean and well aerated. When the fertilized eggs gradually change from orange to dark grey in colour, hatching is imminent, and 5% sea water should be added in succession. The eggs generally require 19 to 20 days to hatch at water temperature of 26° to 28°C.

After all the eggs are hatched, the spawner is removed. More sea water is added to increase the salinity to about 13‰. Some green water is also

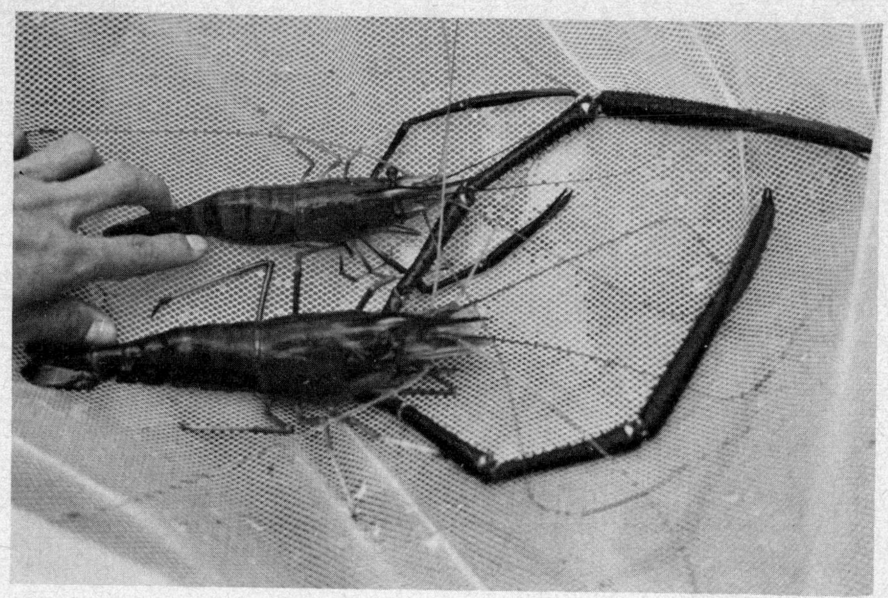

Fig 97 Giant freshwater prawn *Macrobrachium rosenbergii* (female top, male bottom)

Fig 98 From single male and female breeders come hundreds of thousands of offspring

added. From the first to fifth day, thay are given Artemia nauplii and a prepared feed, daily addition of sea water is made to increase the salinity to 13–17‰. In the following days the salinity is further incressed to 18‰, and more mixed feed (flesh of fish and shellfish, egg custard, etc) is given.

At temperatures of 26° to 28°C, the larvae metamorphose into juveniles, and the water is gradually freshened. The juveniles at this stage are about 7 mm in length and may be transferred to rearing ponds (cement ponds with mud bottoms). They should be reared to about 5 cm in length before stocking in the growers pond.

The growers pond is usually a mud pond. Feeds consist of trash fish, shellfish, other animal matters and feeds of plant origin.

8. VIVIPAROUS SNAILS[5]

The viviparous snail, *Viviporus chinensis malleatus,* is a new comer in the aquacultural industry. Its culture is on a small scale, and only a few small farms can be found on the island.

It is a temperate species, with most suitable temperature for growth from 20° to 26°C. It does not feed at temperatures below 15°C and above 30°C.

They are generally reared in ditches of about 1·5 m in width so that they could be picked up by stretching out the hand. The depth of the ditch is 50–60 cm. Flowing water enters the ditch from one end and leaves from the other end. The water inlet and outlet are screened to prevent escape. Water level is maintained at 30 cm. Flowing water is needed because the viviparous snails are sensitive to low oxygen content.

The ponds are manured with compost and left for two weeks before stocking. The season of propagation is from March to October. Fertilization is internal. The tiny baby snails weigh only 0·063 g each. With a sufficient flow of water, 3 to 5 kg of snails may be stocked in each 3·5 m^2 of area.

The viviparous snail is omnivorous, feeding on any plant and animal matters. The farmers usually feed them with fish waste, unused vegetables, etc once every 3 or 4 days. The quantity given is 1 to 3% of the weight of the snails.

They reach 12 to 15 g in individual weight in a year and may be marketed. They are caught by hand, sometimes with the help of a dip net. They are popular in food stalls as well as in homes.

LITERATURE CITED

1. Chen, Gin-fu: Induced Spawning of the Catfish, *Parasilurus asotus* (Linnaeus). *JCRR Fisheries Series* No. 12 1972.
2. Liu, Chia-kang: Notes on Experimental Rearing of Swamp Eel. *JCRR Fisheries Series* No. 11. 1971.
3. Ling, S. W.: The General Biology and Development of *Macrobrachium rosenbergii* (de Man). *FAO Fish. Rep.*, Vol. 3, 1969.
4. Liao, I-chiu, Nai-hsien Chao and Lung-sung Hsieh: Preliminary Report of the Experiments on the Propagation of the Giant Freshwater Prawn, *Macrobrachium rosenbergii*, in Taiwan. *Journal of the Fisheries Society of Taiwan*, Vol. 2, No. 2 1973.
5. Teng, H. T. *et al*: Newly Cultured Freshwater Species. *Taiwan Fisheries Research Institute, Fish Culture Circular* No. 42. 1971.

List of other books published by Fishing News Books Limited

Free catalogue available on request

A living from lobsters
Better angling with simple science
British freshwater fishes
Coastal aquaculture in the Indo-Pacific region
Commercial fishing methods
Control of fish quality
Culture of bivalve molluscs
Eel capture, culture, processing and marketing
Eel culture
Escape to sea
European inland water fish: a multilingual catalogue
FAO catalogue of fishing gear designs
FAO catalogue of small scale fishing gear
FAO investigates ferro-cement fishing craft
Farming the edge of the sea
Fish and shellfish farming in coastal waters
Fish catching methods of the world
Fish farming international 1 and 2
Fish inspection and quality control
Fisheries oceanography
Fishery products
Fishing boats of the world 1
Fishing boats of the world 2
Fishing boats of the world 3
Fishing ports and markets
Fishing with electricity
Freezing and irradiation of fish
Handbook of trout and salmon diseases
Handy medical guide for seafarers